Engines and Electrics

(381 part 2 including Applied Studies)

Editor Roy Brooks

Name. . . . .

College . . . .

Employer . . . .

Session		Class	Ref. No.	Session		Class	Ref. No.
Day and Time	Room	Subject	Teacher	Day and Time	Room	Subject	Teacher
. . . .	. . . . . .	. . . .	. . . .	. . . .	. . . . . .	. . . .	. . . .
. . . .	. . . . . .	. . . .	. . . .	. . . .	. . . . . .	. . . .	. . . .
. . . .	. . . . . .	. . . .	. . . .	. . . .	. . . . . .	. . . .	. . . .
. . . .	. . . . . .	. . . .	. . . .	. . . .	. . . . . .	. . . .	. . . .
. . . .	. . . . . .	. . . .	. . . .	. . . .	. . . . . .	. . . .	. . . .
. . . .	. . . . . .	. . . .	. . . .	. . . .	. . . . . .	. . . .	. . . .

Macmillan Motor Vehicle Craft Studies Series

Engines and Electrics

(381 part 2 including Applied Studies)

Editor   Roy Brooks
            Senior Lecturer in Motor Vehicle Subjects
            Bolton Institute of Higher Education

Author   Jack Hirst
            Lecturer in Motor Vehicle Subjects
            Burnley College of Arts and Technology
            Examiner in Motor Vehicle Subjects to the
            West Midlands Advisory Council for Further Education
            incorporating Union of Educational Institutions

MACMILLAN
EDUCATION

First published 1982
Reprinted 1985, 1986, 1987, 1988, 1989, 1990

Published by
MACMILLAN EDUCATION LTD
Houndmills, Basingstoke, Hampshire RG21 2XS
and London
Companies and representatives
throughout the world

Printed in Great Britain by
Vine and Gorfin Ltd
Exmouth

ISBN 0–333–34643–2

Contents

The contents follow, as far as practicable, the order of the City and Guilds syllabus. This may not always be the best teaching/learning sequence, but it ensures a uniformly understandable order of presentation.

The publishers will be happy to suggest an effective scheme of work based on page numbers.

PREFACE

This book deals with the engines and vehicle electrical components as required by syllabus 381 part 2 Light and Heavy Vehicle Mechanics, 'Engines and Electrics'. The author has wide experience in these fields and was primarily responsible for covering the same topic areas in the earlier highly successful books in the Macmillan Motor Vehicle Craft Studies Series. He is also a well-known and respected examiner.

As before, the requirements of the City and Guilds syllabus have been faithfully followed, and the book's aim is to save both student and teacher time in coping with the demands of a necessarily full syllabus.

Obviously, the maximum benefit will be obtained from this book if it is used in conjunction with its companion volume, *Transmission, Chassis and Materials*, by John Whipp. Anyone conscientiously completing these two books should easily pass the examinations in these topic areas.

The book is suitable for both Light and Heavy vehicle mechanics, since substantial areas of their work overlap. Where divergences do occur, such sections can be omitted — each section has a precise syllabus reference in the margin.

While wishing every success to those who use the books in this series, the authors and editor realise that improvements are always possible. Constructive comments are welcome. Please send them via the publishers — they will be most carefully considered.

Roy Brooks

ACKNOWLEDGEMENTS

AC Delco Division of General Motors Ltd
Audi NSU Auto Union AG
BL Ltd
Robert Bosch Ltd
The British Petroleum Company Ltd
The British Standards Institution
CAV Ltd
Champion Sparking Plug Company Ltd
Crypton-Triangle Ltd
Cummins Engine Company Ltd
Datsun UK Ltd
Alexander Duckham & Company Ltd
Fiat Auto (UK) Ltd
Ford Motor Company Ltd
General Motors Ltd
Halls Gaskets Ltd
Leslie Hartridge Ltd
Hepworth & Grandage Ltd
Holset Engineering Company Ltd

Honda UK Ltd
Lotus Group of Companies
Joseph Lucas Ltd
Lumenition Ltd
Mobelec Ltd
Mobil Oil Company Ltd
Perkins Engines Ltd
Renault (UK) Ltd
Ripaults Ltd
Smiths Industries Ltd
Solex UK Ltd
SU Butec
Suntester Ltd
Talbot Motor Company Ltd
VAG (United Kingdom) Ltd
Vauxhall Motors Ltd
Villiers
Weber Carburettors UK Ltd

The motor vehicle staff of Burnley college

ENGINES – TECHNOLOGY

PRINCIPLES OF ENGINE CONSTRUCTION

The exploded view shows most of
the main engine structure and
moving parts used in a four-cylinder
in-line engine with overhead
camshaft.

Observe how the components
would fit together to form the
complete engine.

Study the interrelationship of all
the moving parts, how each
component is dependent upon
others for its operation and how
they contribute to the overall
design.

It is expected that you can identify the main components on these drawings from previous experience.

The numbered items are small but essential parts of the engine.

Name these components.

No.	Name
1.	....
2.	....
3.	....
4.	....
5.	....
6.	....
7.	....
8.	....
9.	....
10.	....
11.	....
12.	....
13.	....
14.	....
15.	....
16.	....
17.	....
18.	....
19.	....
20.	....

Note. The remainder of this chapter will cover the variations in design of the many engine components.

It is expected that these two pages will be constantly referred to while completing this work.

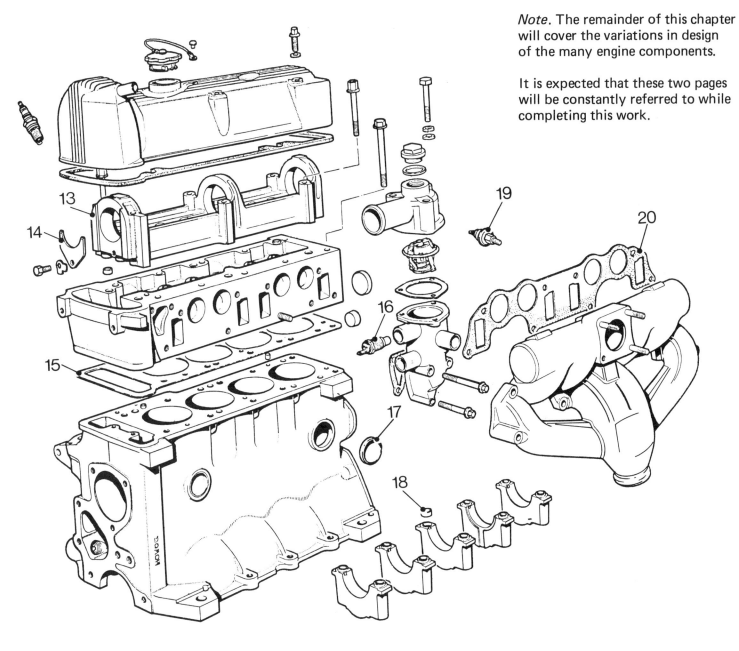

FOUR-STROKE CYCLES

Describe, emphasising their basic differences, the four-stroke, spark-ignition and compression-ignition engine operating cycles. Mention typical timings and compression ratios.

Spark-ignition

Induction

....
....
....
....

Compression

....
....
....
....
....

Power

....
....
....
....
....
....

Exhaust

....
....
....

inlet valve — spark plug — exhaust valve — piston — connecting rod — crankshaft

Compression-ignition

Induction

....
....
....
....

Compression

....
....
....
....
....

Power

....
....
....
....
....

Exhaust

....
....
....

TWO-STROKE CYCLE – COMPRESSION-IGNITION ENGINE

Describe, with the aid of the diagrams provided, the cycle of operations during the two strokes.

First stroke

....
....
....
....
....
....
....
....
....
....

Second stroke

....
....
....
....
....
....
....
....
....

Why is it necessary to pressure-charge this type of engine?

....
....
....
....
....
....
....
....
....
....
....

List three advantages gained by using this type of pressure-charged engine.

....
....
....
....
....

5

GAS TURBINE

Gas turbine engines, mainly of the two-shaft design, are used in a few heavy commercial vehicles and certain other specialised applications. With the aid of the simplified schematic diagram below, explain the principle of operation. Add arrows to the diagram to indicate direction of gas flow.

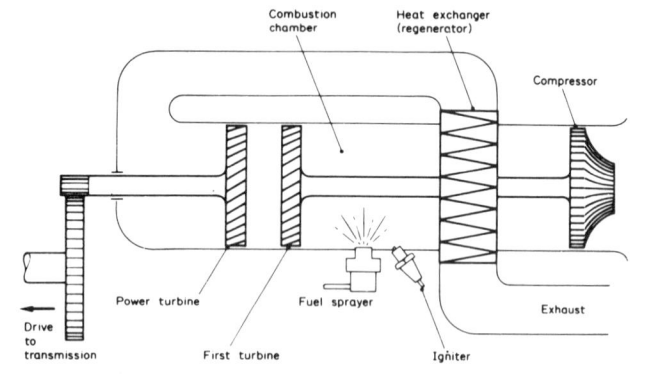

Note. The compressor turbine shaft is initially operated by a drive from the starter motor.

....
....
....
....
....
....
....
....
....
....
....

Answers to the following questions need not relate to any specific engine but should be typical of gas turbines in general.

What type of fuel is commonly used? ....

In what way does the method of ignition and fuel supply differ from both the petrol and C.I. engine?

....
....
....
....

To what pressure is the air usually compressed? ....

How is the combustion chamber cooled?

....
....

Quote typical engine speeds for idling and maximum revolutions.

Idling .... Maximum ....

What would be the expected engine life as compared with a piston engine of similar power?

....
....
....

How does the torque output differ when compared with a piston engine?

....
....
....
....

Opposite are shown two simplified line drawings of a modern gas turbine engine

Note. The method of turning the engine to start and the regenerator drives are not shown.

Name the important parts and indicate with arrows the flow of gas through the system.

Describe this gas flow through the system.

....
....
....
....
....
....
....
....

What are the advantages and disadvantages of the gas turbine as a vehicle power unit as compared with the conventional piston engine?

ADVANTAGES

....
....
....

DISADVANTAGES

....
....
....

Drawing A shows a sectioned side-view in which the turbines, adjustable turbine blades and combustion chamber can be clearly seen.

A

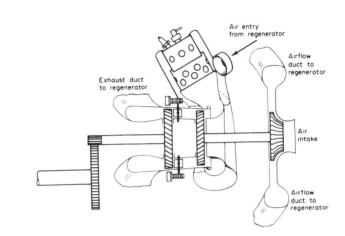

B

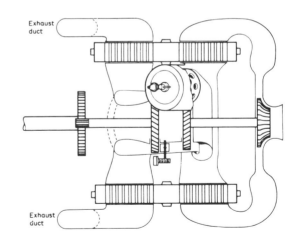

Drawing B is a plan view which also shows the position of the two heat regenerating discs.

7

ROTARY (WANKEL) ENGINE

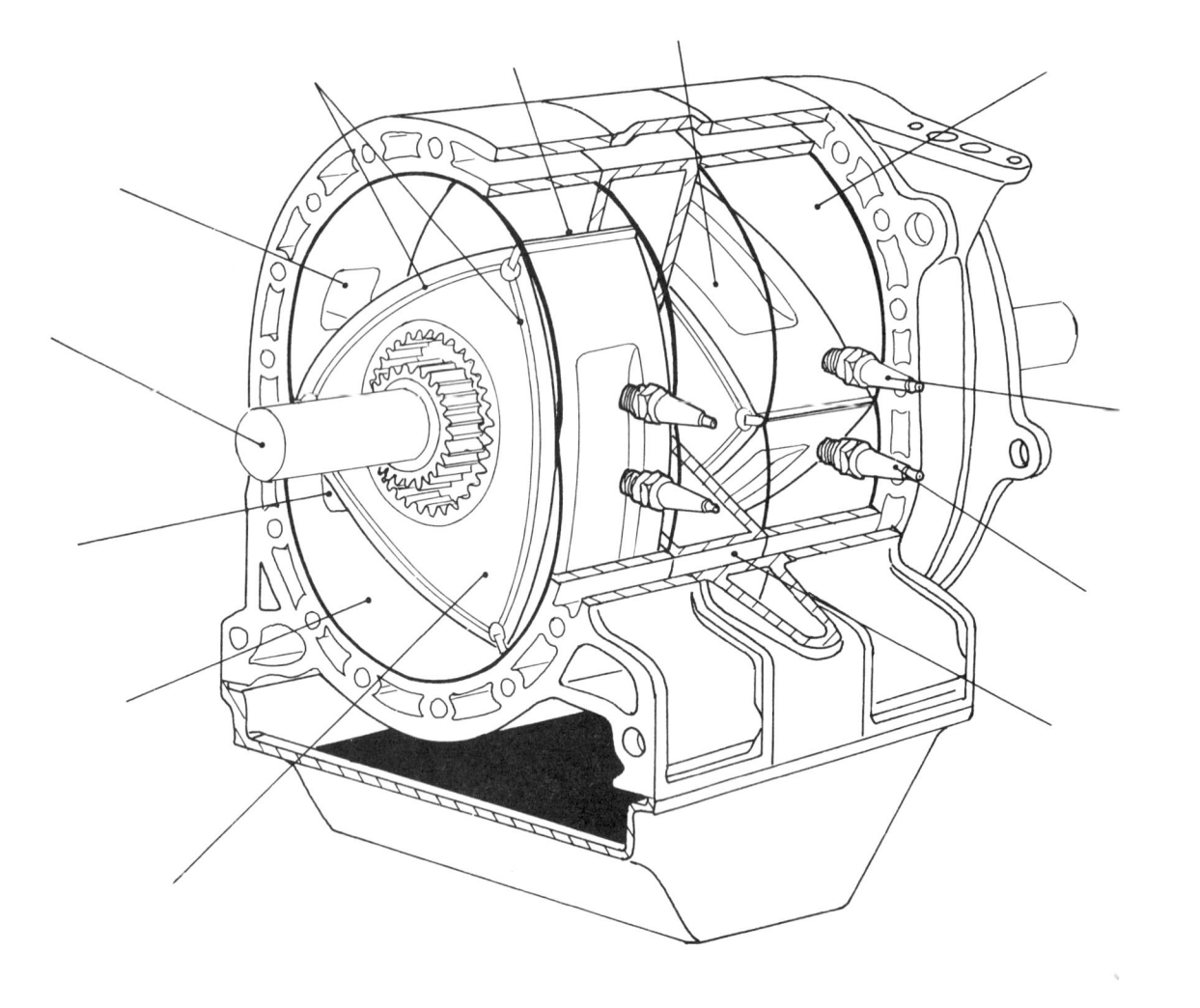

The diagram shows one type of rotary engine; note the use of two spark plugs in each 'cylinder' in this particular application.

Name the main parts indicated on the drawing.

What advantages are to be gained by using this type of engine as against a conventional engine of similar power?

....
....
....
....
....

What are the main reasons why this type of engine has not been brought into widespread use?

....
....
....
....
....
....
....
....
....
....

ROTARY (WANKEL) ENGINE

This type of engine reproduces the four-stroke cycle by the turning action of the equilateral, triangular-shaped rotor in the epitrochoidal bore.

The rotor turns at one-third the speed of the crankshaft. Prove this from observing the drawings.

....

....

....

....

With reference to the diagrams below explain the operational sequence of events.

How many firing pulses occur during each complete rotor revolution?

....

How many power strokes occur during one crankshaft revolution?

....

If the engine speed is 3000 rev/min, what is the speed of the rotor?

....

The rotor is turning clockwise.

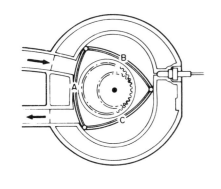 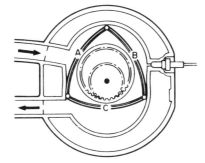 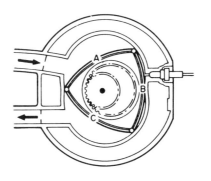 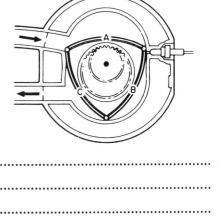

....

....

....

....

....

....

....

....

CAMSHAFTS

L1:6,7
H1:5,6

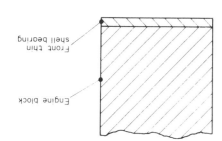

The function of the camshaft is to operate the valves, and frequently it forms a convenient mounting point for various auxiliary drives. Its position on the engine and the actual shape of the cams themselves vary according to individual manufacturers' requirements.

Name three auxiliaries that could be driven by the camshaft.

....

Suggest one important advantage of a side-mounted camshaft as compared with an o.h.c. layout.

....

What features dictate cam shape?

....
....

State the method used in the above sketch to locate the drive gear on the shaft.

....
....

Suggest an alternative method of accurate location of the cam drive gear on the shaft.

....

Complete the diagram below to show a method of controlling camshaft end-float, also show how the drive gear is located on the shaft.

Indicate the end-float clearance stating a typical value.

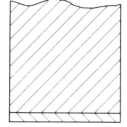

Engine block

Front thin shell bearing

Describe how the end-float can be controlled (altered).

....
....
....

CAMSHAFT DRIVES

The camshaft must be driven from the crankshaft. This may be done by gear, chain or toothed belt.

Give below, details of the relative advantages and disadvantages.

Advantages	Disadvantages
Gear drive	**Gear drive**
....	....
....	....
....	....
....	....
....	....
Chain drive	**Chain Drive**
....	....
....	....
....	....
....	....
....	....
Toothed belt-drive	**Toothed belt-drive**
....	....
....	....
....	....
....	....
....	....

Show the drive gear arrangements when the camshaft is mounted in the cylinder block and driven by:

(a) chain — indicate the direction of rotation and show clearly the tensioner arrangement.

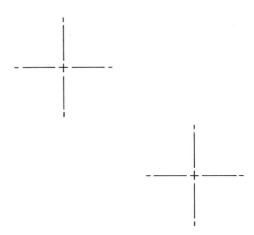

(b) gears only — indicate direction of rotation and state the material from which the gears may be made.

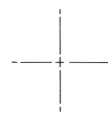

11

Sketch two methods of driving a single overhead camshaft.

Show method of tensioning.

1. Vehicle

   Make ....

   Model ....

2. Vehicle

   Make ....

   Model ....

Sketch a method of driving a twin over-head camshaft engine.

Vehicle

Make ....

Model ....

Sketch (below) an arrangement that includes driving the fuel injection pump on a C.I. engine as well as the camshaft.
Vehicle

Make ....

Model ....

12

TENSIONERS

On side-camshaft drive engines the simplest type of tensioner is a spring-loaded, metal plate normally faced with bonded rubber.
(Typical such arrangements should have been adequately shown in the drawings on the previous pages.)

Automatic tensioners

These types usually have some form of ratchet which ensures that a constant spring pressure is maintained on the chain.

Oil-pressure-assisted tensioner

Describe the action of the tensioner shown.

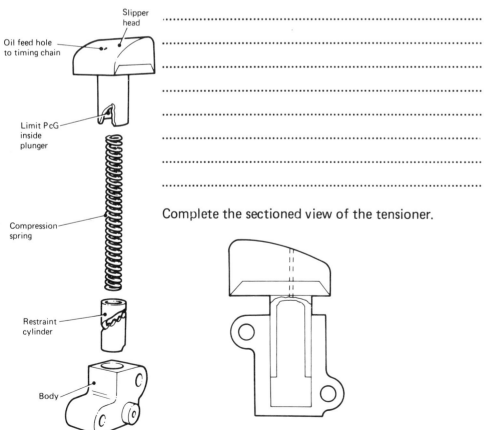

Slipper head

Oil feed hole to timing chain

Limit PcG inside plunger

Compression spring

Restraint cylinder

Body

....
....
....
....
....
....
....
....

Complete the sectioned view of the tensioner.

AUTOMATIC MECHANICAL TENSIONER

Snail cam tensioner

Spring

Synthetic rubber pad

Snail cam

Tensioner arm

Pivot

Describe how this type maintains tension.

....
....
....
....
....
....
....

Belt-drive tensioner

Belt stretcher and spring
Belt stretcher hole

Stretcher lock nut

Jockey pulley

Stretcher pivot nut

Crankshaft pulley

The type shown is more commonly known as a jockey pulley. The pulley rotates on the outside of the belt.

Describe how the belt is tensioned.

....
....
....
....
....
....

LUBRICATION

Describe the most common method of lubricating

(i) camshaft bearings ....
....

(ii) cams ....
....

13

PUSH RODS AND ROCKERS (VALVE OPERATION)

On overhead valve engines with the camshaft positioned in the cylinder block the valves are operated via cam followers, tappets, push rods and rockers.

On some modern push-rod-operated engines the rocker shaft has been eliminated.

Examine such a mechanism and complete the diagram below.

Complete the diagram to show a rocker fixed to a rocker shaft. Indicate the method of adjustment and lubrication.

The length of the push rod depends upon the camshaft position in the block.

The tappets are usually hollow and flat-bottomed. The underside is heat treated to resist wear.

Show how the cam can be positioned to allow the tappet to rotate. Why is this desirable?

....
....
....
....
....
....
....
....
....

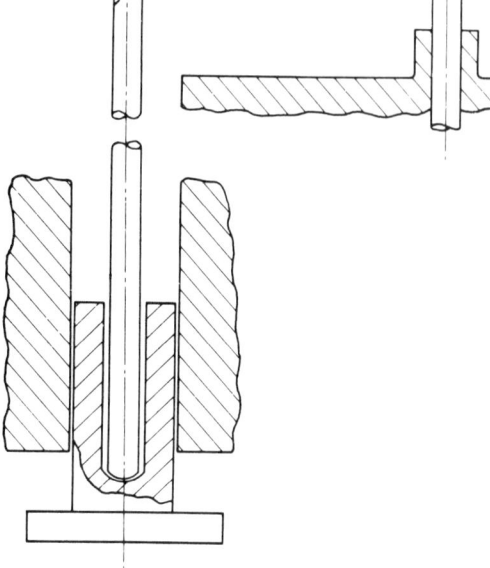

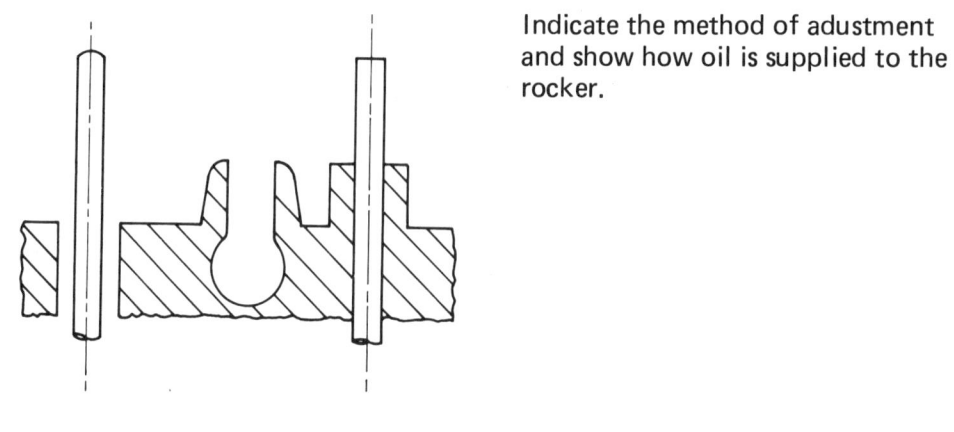

Indicate the method of adustment and show how oil is supplied to the rocker.

What is the advantage of using this type of mechanism?

....
....
....
....

Why do some manufacturers recommend that the engine should be running when the valve clearance is adjusted?

....
....
....

14

OVERHEAD CAMSHAFT VALVE ARRANGEMENTS

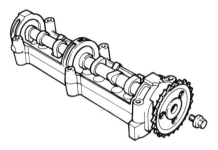

With the overhead camshaft design the most common arrangement is to allow the cam to operate directly on the tappet block which is in direct contact with the valve.

Draw the adjusting shim in its correct position.

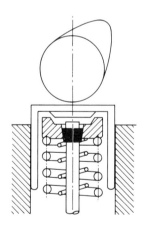

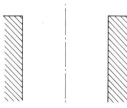

The above design shows a method of obtaining the valve clearance by using shims of varying thickness. *Note.* When the engine has been assembled the clearance cannot be adjusted.

Complete this diagram to show a design where it is not necessary to remove the camshaft when adjusting valve clearance.

An alternative arrangement is to have the cam operate the valve via finger levers.

Give an advantage and disadvantage of this arrangement compared with the other two arrangements on this page.

....

....

....

Indicate the provision for valve adjustment.

Show the valve stem oil seal on the above drawing.

What are the main advantages of using an o.h.c. design compared with side camshaft?

....

....

....

....

....

15

Examine engines that operate on the spark-ignition and compression-ignition four-stroke cycle.

Compare the main engines and auxiliary components and state any constructional differences that are created by the different principles of operation.

The first line is completed as an example.

Items examined	Spark-ignition	Compression-ignition
Combustion chamber	Formed in cylinder head	Formed in piston crown ....
Pistons	....	....
Crankshaft	....	....
Fuel supply	....	....
Air intake	....	....
Auxiliaries	....	....

POPPET VALVES

In almost all four-stroke engines the poppet valve controls the flow of gas into and out of the cylinder.

Complete the diagrams below to show valve heads of an alternative shape to the flat head.

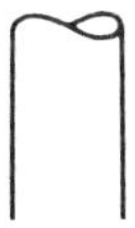

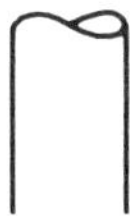

Type ....   Type ....

....

....

On modern vehicles the valve seat angle usually is ....

On some C.I. engines the inlet valve-seat angle may be ....

The diagram shows the correct contact of valve to seat.

Shade on the valve the area of valve-seat contact.

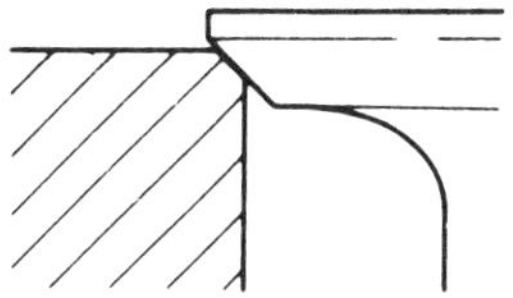

State the effects of

(i) excessive clearance ....

....

....

(ii) insufficient clearance ....

....

....

VALVE SPRING RETENTION

The most common arrangement is by split collets.

Remove a valve from a cylinder head. Examine it and complete the sketch opposite. Show clearly the method of spring attachment and stem oil-sealing arrangement.

....

....

....

....

....

....

....

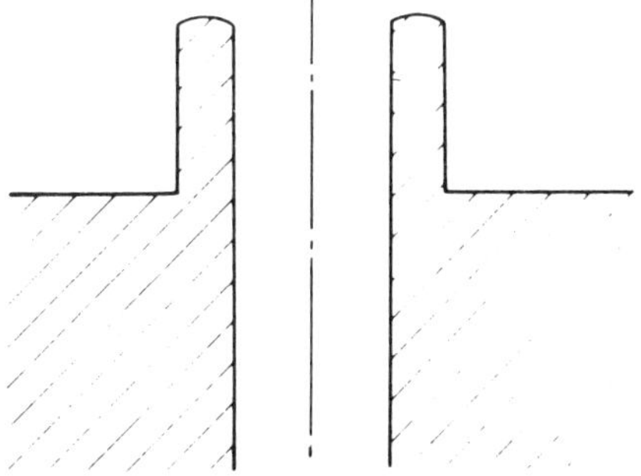

Sketch two other alternative types of collar attachments.

Type of attachment ....   Type of attachment ....

17

VALVE GUIDES

The valve stem is located in a guide which may be integral with the head or in the form of a sleeve which is an interference fit in the head or block.

At (b) and (c) immediately below, sketch two types of interference fit valve guides and name each type of material.

(a)　　　　　　　　(b)　　　　　　　　(c)

Type ...*Integral*...　　Type ....　　Type ....

Material ...*Cast iron*...　　Material ....　　Material ....

How would excessive clearance caused by wear be brought back to standard in type:

(a) ....

....

(b and c) ....

....

State a typical valve stem clearance ....

What would be the effect of:

Insufficient stem clearance ....

Excessive stem clearance ....

....

VALVE SPRINGS

The valve is opened, either directly or indirectly, by a cam on the camshaft and is usually closed by a coil spring.

The valve spring may be a single spring having either (a) uniform coils, or (b) the coils wound closer together at the cylinder head end or (c) two springs, one inside the other.

Name a vehicle where each type above is used.

(a) .... (b) .... (c) ....

Complete the sketches to show:
Two springs, one inside the other.

Spring having closer coils at the lower end.

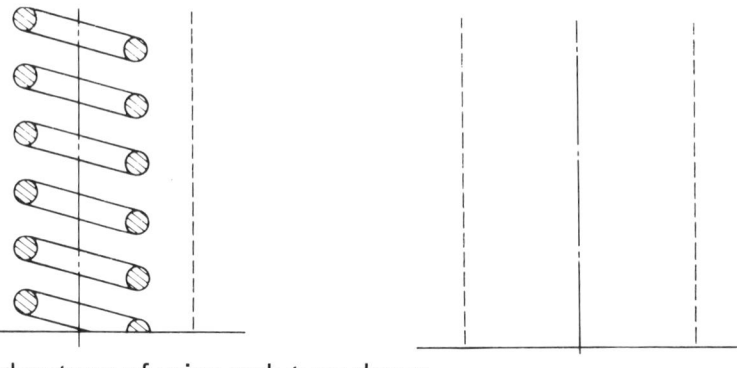

Give the advantages of using each type shown.

....　...

....　...

....　...

What is meant by:
valve bounce? ....

....

....

VALVE-OPENING PERIODS

The valves in an engine do not open and close exactly at the beginning and end of their respective strokes. What is the reason for the extension of their open period?

....

....

....

....

Extending the valves open period will improve the volumetric efficiency of the engine. What is meant by this term volumetric efficiency?

....

....

....

....

Why does altering the valve timing affect the volumetric efficiency?

....

....

....

....

....

State some other functions that can improve the volumetric efficiency of the engine.

....

....

....

....

What is meant by the following and when do they occur?

Valve overlap ....

....

....

Valve lead ....

....

....

Valve lag ....

....

....

Indicate on the diagram examples of where lead, lag and overlap occur.

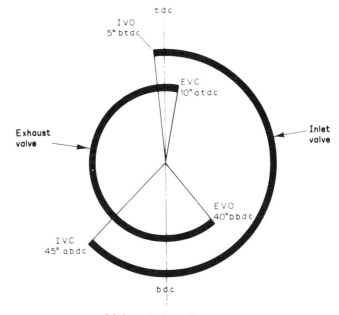

Valve-timing diagram

COMBUSTION SPARK-IGNITION

The air:fuel mixture is compressed and just before the piston reaches top dead centre the spark ignites the mixture.

Describe the combustion process that then occurs.

....

....

....

....

With regard to the combustion chambers, state what features help to achieve the following:

1. High volumetric efficiency.

....

....

....

2. Short flame travel path.

....

....

3. Maximum heat transfer from hot regions to the cooling air or water.

....

....

4. Suitable quench regions at the more distant parts of the combustion chamber, i.e. furthest away from spark plug.

....

....

....

5. Correct degree of turbulence of the compressed mixture.

....

....

6. Good scavenging of the exhaust gases.

....

....

7. Adequate cooling of the exhaust valve head and spark plug points.

....

....

....

....

To obtain maximum thermal efficiency (minimum heat loss) the design of the combustion chamber and its associate components must achieve, without the risk of detonation, the highest compression ratio possible when using fuel of a given octane rating.

What is the main problem when compression ratios are raised?

....

....

State ways in which the compression ratio may be varied and the effect on the performance of the engine.

....

....

....

....

20

TYPES OF COMBUSTION CHAMBERS USED IN SPARK-IGNITION ENGINES

Complete each side and plan view, in line-diagram form, to show the positions of the valves and combustion-chamber shapes.

State the features that make these types of combustion chambers suitable for spark-ignition engines.

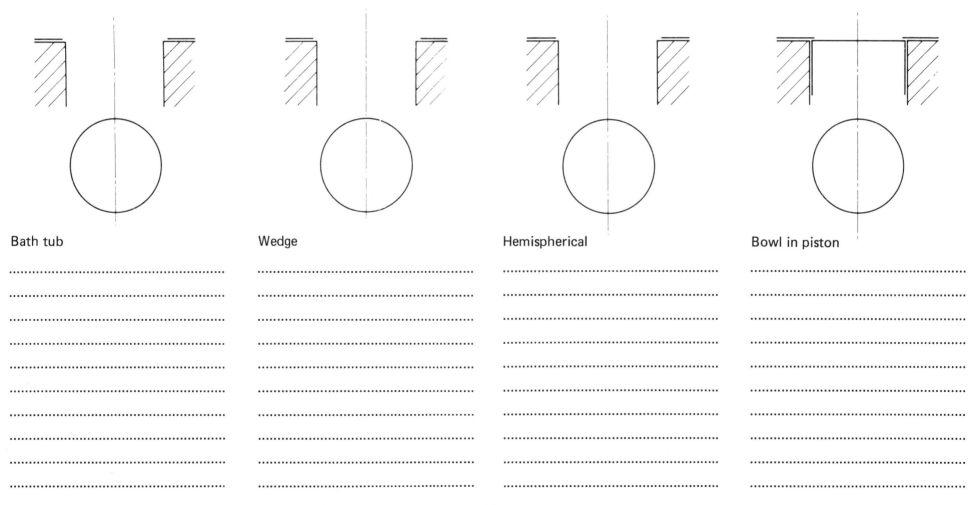

Bath tub

Wedge

Hemispherical

Bowl in piston

COMBUSTION IN A COMPRESSION-IGNITION ENGINE

The combustion process in a compression-ignition engine is said to have three distinct stages or phases.

First phase   Delay period
Second phase   Rapid pressure rise
Third phase   Continuous burning

Describe the combustion process through these phases.

....
....
....
....
....
....
....
....
....
....
....
....

See 1:74 for graph showing the three phases.

What factors affect the length of the delay period?

....
....
....

DIRECT INJECTION

In the first diagram draw a sectioned view of the type of piston used in a direct injection engine.

Show on both diagrams the air and fuel swirl as the piston is nearing top dead centre.

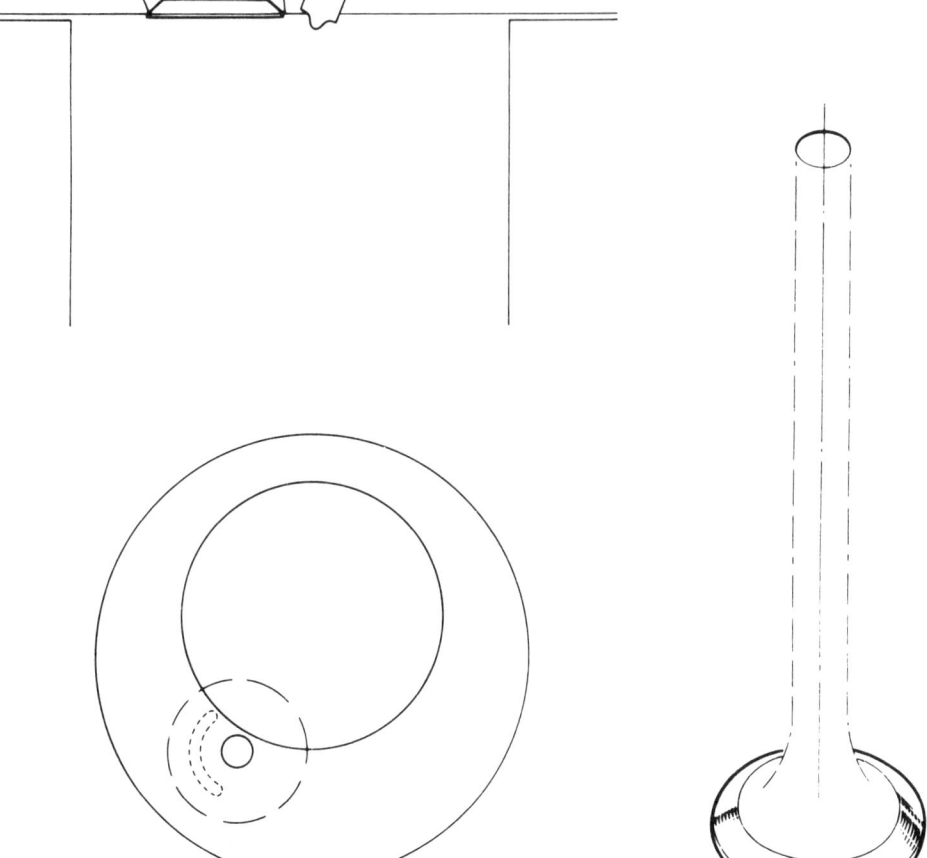

Draw on the valves shown (*left and below*) the mask or shroud; also on the large valve below indicate how it is prevented from rotating.

22

DIRECT INJECTION

Describe the important features of the basic layout shown on the previous page and explain how turbulence and swirl are induced and controlled.

....

....

....

....

....

INDIRECT INJECTION

Describe the important features of the basic indirect injection layout opposite.

Explain how the turbulence is created in the two pre-combination chamber designs you have shown.

....

....

....

....

....

State the main advantages of using direct injection. ....

....

indirect injection. ....

....

What size-range of engine most commonly uses

direct injection? ....   indirect injection? ....

Sketch two different methods of indirect injection design (e.g. Perkins and Ricardo).

Cylinder block

Engine make ....   Model ....

Cylinder block

Engine make ....   Model ....

23

CYLINDER LINERS

Cylinder liners are fitted into some engine blocks as shown opposite.

What purpose do they serve?

....

....

....

State two types of cylinder liner material:

1. ....

2. ....

List the main advantages of fitting cylinder liners.

....

....

....

What are the main causes of rapid cylinder bore wear?

....

....

....

What reasons may cause substantial variations in wear between different cylinders on the same engine?

....

....

....

....

Identify the type of liners shown.

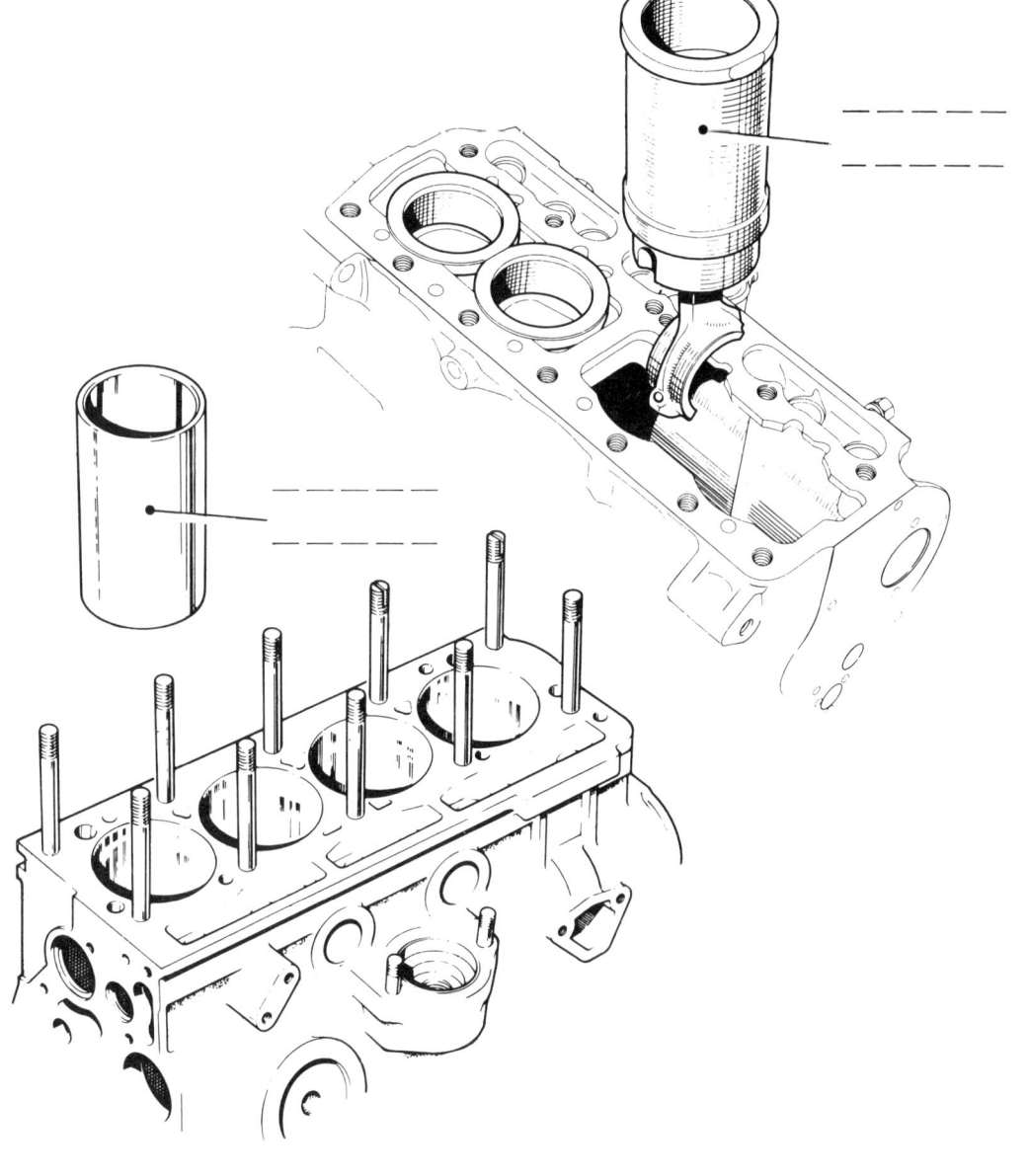

DRY CYLINDER LINERS

These are pressed directly into the cylinder bore.

Two types are used:

Show a sketch of both types fitted into the cylinder block, include the water space.

1. Interference fit

....
....
....
....
....
....
....
....

Interference fit liner

2. Slip fit

....
....
....
....
....
....
....
....

Slip fit liner

WET CYLINDER LINERS

The type of liner has its outer surface in direct contact with the cooling water. There are two types in common use:

(a) The liner's top shoulder is located in a recess in the cylinder block faces whilst the lower end is a push fit into the lower part of the block.

Complete the diagrams below

....
....
....
....
....
....

Type commonly fitted in C.I. engines

(b) This type block has an 'open deck' layout with the liners inserted into spigots rising from the lower part of the block.

....
....
....
....
....
....

Type commonly fitted in o.h.c. engines

25

INVESTIGATION

1. Interference fit liners

Examine an interference fit cylinder liner and the cylinder block it is to fit. If possible measure the cylinder bore diameter and external cylinder liner diameter and determine the interference.

What checks should be made on the cylinder block before fitting the liner?

....

....

What should be done to assist the liner to be fitted into the bore?

....

....

How must the liner be finished off when fitting is complete with regard to:

(a) the cylinder block-head face ....

(b) its internal diameter ....

2. Wet liners

What is the object of providing cylinder liner nip?

....

....

....

When the cylinder head is removed for any reason, what precautions should be observed with regard to this type of liner?

....

....

....

Note the type of fit and sealing arrangements for a wet cylinder liner and block, and determine the amount of protrusion when fitted to provide 'nip'.

Engine make ....   Model ....

Fitting the liner	Comments
1. Examine the locating recess or lip and note condition. Clean if necessary.	....
2. Fit new sealing ring smear with suitable lubricant.	....
3. Push liner into cylinder block.	....
4. Bolt down and check liner protrusion.	....

Actual protrusion ....   Manufacturer's specification ....

Water space holes

Cylinder liner stood proud only by 0.25mm

Show on the sketch a method of holding down the cylinder liner during servicing.

26

CYLINDER BLOCKS

On modern engines the 'cylinder block' is the term usually given to the combination of the block itself which carries the cylinders, and the crankcase, which supports the crankshaft, i.e. the basic 'framework' of the engine.

A separate cylinder block is commonly used for motor-cycle, air-cooled and heavy commercial vehicle engines.

State TWO reasons for adopting this separate cylinder block type of construction for:

Give a more correct definition of the cylinder block.

....

....

....

....

....

Single-cylinder motor-cycle engines

1. ....

....

2. ....

....

Air-cooled multi-cylinder engines

1. ....

....

2. ....

....

Heavy commercial vehicles

1. ....

....

2. ....

....

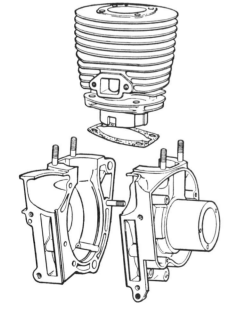

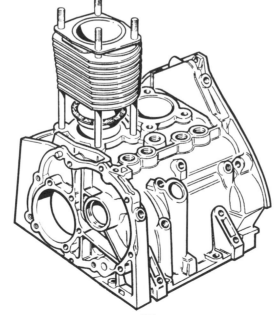

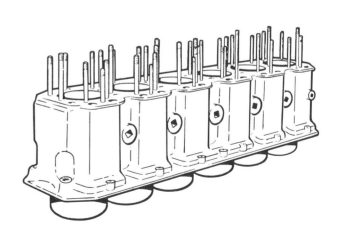

CRANKCASES

The engine crankcase may be a separate unit supporting the crankshaft and camshaft or be an integral part of the cylinder block.

Two basic designs in current use are:

(a) Separate crankcase.

....

....

(b) Combined cylinder block and crankcase, where the crankcase lower face is in line with the centre line of the main bearings, or is extended below the centre line of the main bearings as shown below.

....

....

Identify the important parts of the integral crankcase shown below.

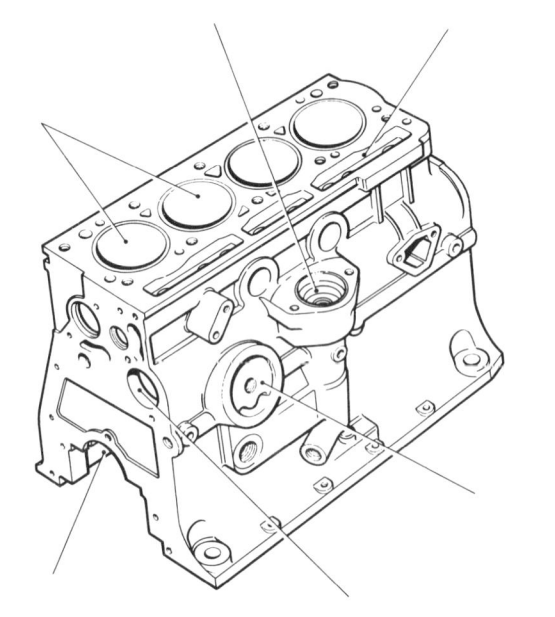

Examine a crankcase and cylinder block similar to the one opposite and complete the diagram below to show a section through one of the cylinders.

Indicate the following items:
engine cylinder
crankcase webbing
camshaft bearing
main bearing
tappet guide
water jacket

How are the effects of distortion and vibration created in the crankshaft minimised?

....

....

....

....

Why are there always core (welch) plugs fitted to the side of the cylinder block of a multi-cylinder water-cooled engine?

....

....

....

....

28

CRANKCASE VENTILATION

Give two reasons why a flow of air through the crankshaft is necessary.

....

....

....

....

....

....

....

POSITIVE CRANKCASE VENTILATION

On modern cars it is usual to employ some form of crankcase ventilation which does not emit unburnt oil fumes to the atmosphere. What is the reason for this practice.

....

There are two systems of positive crankness ventilation:

1. Semi-closed type ....

....

2. Fully closed type ....

....

The diagram shows the method of crankcase ventilation used on many pre-1972 vehicles. Add arrows to indicate flow.

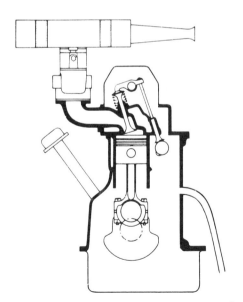

The diagram shows the same engine modified to a positve crankcase ventilation (P.C.V.) system—semi-closed type. Add arrows to indicate ventilation flow.

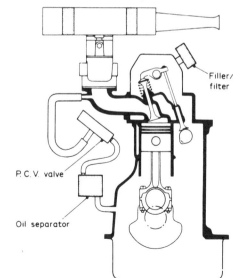

Complete the diagram to show the same engine modified to a P.C.V. system of the fully closed type and add arrows to indicate ventilation flow.

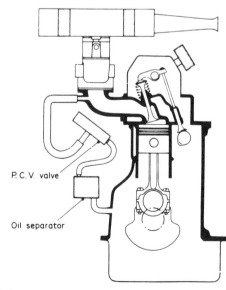

Examine various modern vehicles, note the type of P.C.V. system used and the positioning of the various components.

POSITIVE CRANKCASE VENTILATION

What is the advantage of the fully closed system as compared with the semi-closed system?

....
....
....
....

CRANKCASE VENTILATION VALVE

Two types of valves are widely used, they are both controlled by manifold vacuum which moves either a diaphragm or a plunger (as shown opposite).

What is the ventilation valve's function?

....
....
....
....

What disadvantages may the valve cause?

....
....
....
....

How, and after what period, should these valves be serviced?

....
....

Explain, with the aid of the diagram below, the operation of the A.C. crankcase ventilation valve, giving reasons for the flow and no-flow positions.

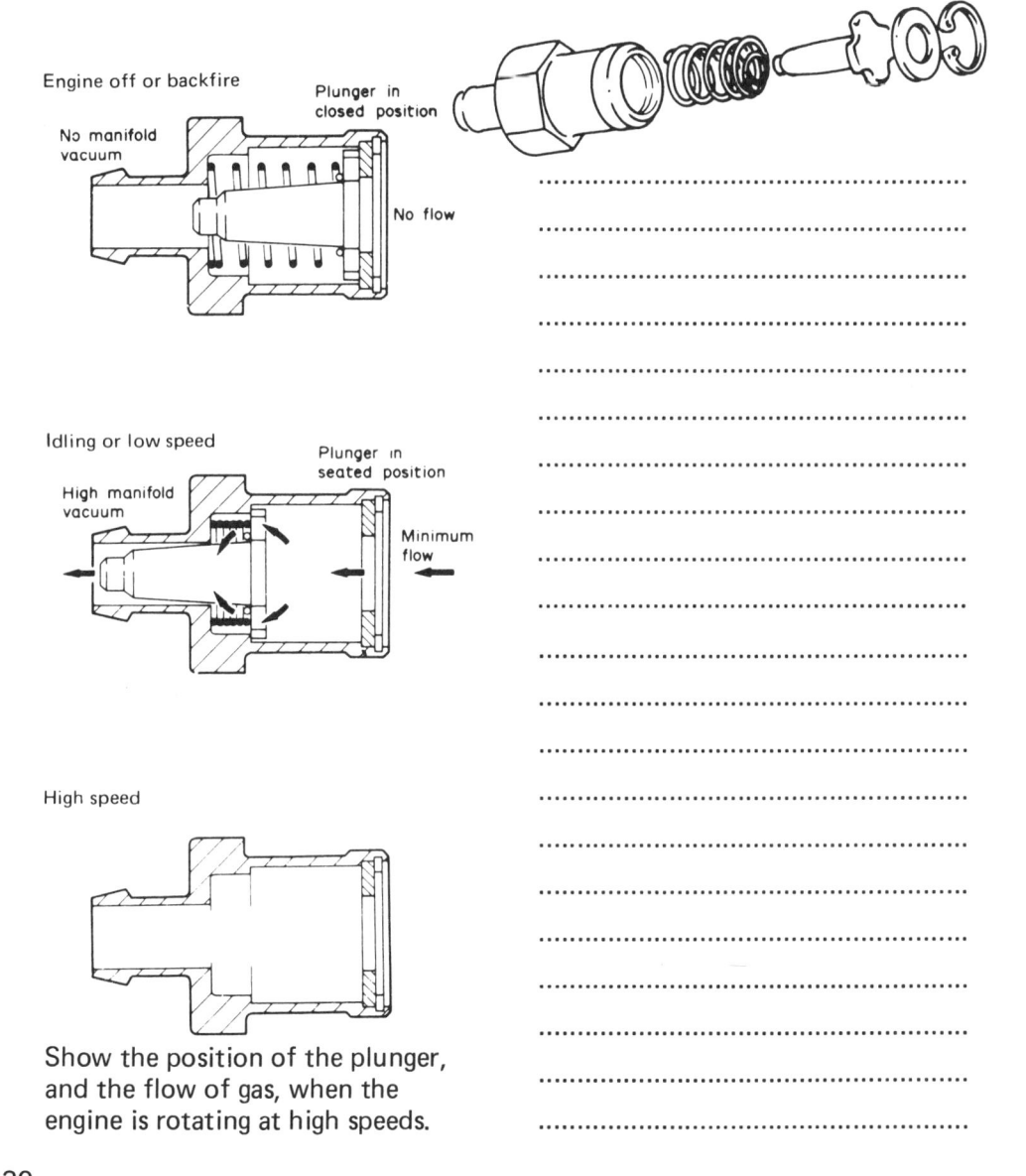

Engine off or backfire

No manifold vacuum

Plunger in closed position

No flow

Idling or low speed

High manifold vacuum

Plunger in seated position

Minimum flow

High speed

Show the position of the plunger, and the flow of gas, when the engine is rotating at high speeds.

....
....
....
....
....
....
....
....
....
....
....
....
....
....
....
....
....
....

CYLINDER HEADS

The cylinder head is mounted on top of the cylinders. It confines the pressure of combustion and directs it down onto the piston. The head also provides water-cooling passages, inlet and exhaust gas passages and supports the valve gear.

Identify the important features of the cylinder heads shown.

Uniflow head

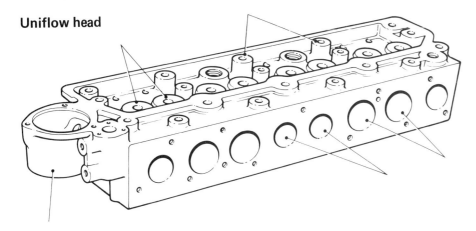

Crossflow head

....

....

....

....

NUT-TIGHTENING SEQUENCES

Cylinder head bolts (or nuts) must be tightened in the correct sequence and to the correct torque. What might be the results if this procedure is not followed?

....

State next to the leader lines a correct tightening sequence.

Light vehicle cylinder head

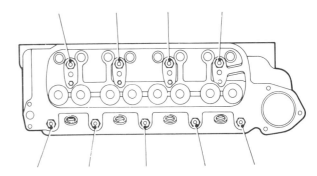

Typical torque setting would be ....

Heavy vehicle cylinder head

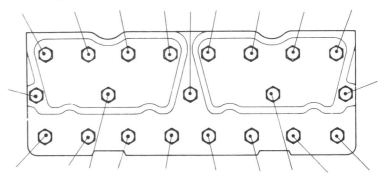

Typical torque setting would be ....

31

CYLINDER HEAD GASKETS

Identify the hole uses.

State the purpose of a cylinder-head gasket.

....
....
....
....

Give some of the main symptoms and causes of cylinder-head gasket failure.

Symptoms

....
....
....

Causes

....
....
....

PISTONS

The piston forms a sliding gas-tight (and almost oil-tight) seal in the cylinder bore and transmits the force of the gas pressure to the small end of the connecting rod. In achieving this the piston must form a bearing support for the gudgeon pin and take side thrust loads created by the angular displacement of the connecting rod.

Name the various parts of the piston below and indicate using arrows the main side thrust loads created by the connecting rod.

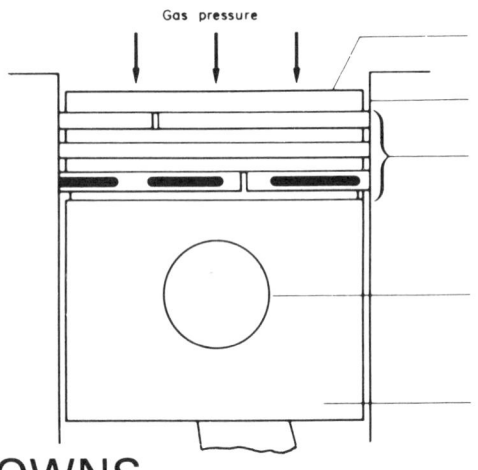

PISTON CROWNS

On certain engines the piston crown (other than simply being flat) is designed to a specific shape. This shape may provide the combustion chamber space and/or improve the gas flow.

Examine various types of pistons and complete the drawings below to show two other special shapes of crown.

Type ....

PISTON CONSTRUCTION

The shape, size, strength and material of a piston are largely dependent upon the type of engine in which it is to be used, e.g. spark-ignition, compression-ignition, type of combustion chamber, etc.

The basic physical differences are covered on p.16 and the drawings may well remind you of these features.

 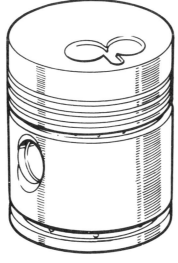

Type ....   Type ....

List the major points that must be considered when designing pistons.

....

....

....

....

....

....

....

GUDGEON-PIN BOSSES

The bosses must be of sufficient cross-section to withstand and transmit the pressures of combustion to the connecting rod. If the support is insufficient the ring belt is likely to deform telescopically, this tends to close the ring grooves and cause piston ring sticking.

Examine a solid-skirt aluminium-alloy sectioned piston and show on the diagram opposite, the shape and position of the gudgeon-pin boss webbing.

....

....

....

....

SKIRT SHAPE

Solid skirt pistons are used in C.I. and high-powered petrol engines, this ensures that the skirt will withstand the heavy loads placed upon it.

Why are these types of pistons often ground slightly oval?

....

....

....

What is meant by a 'slipper'-design piston?

....

....

....

....

THERMAL EXPANSION

In all types of engines it is essential that the operating clearance of the piston in the cylinder is kept to a minimum.

It is necessary to design pistons with operating clearances that can cope with the very high temperatures applied to the crown and its progressive reduction down the piston's length. This creates a considerable thermal expansion variation.

....
....
....
....

1. **Solid skirt** as shown on the previous page.

....
....
....

The diagram shows how the temperature gradients vary on the crown and sides of a typical C.I. engine solid skirt piston.

....
....
....
....
....
....

2. Thermal slot — translot or 'W' design

....
....
....
....
....
....
....
....
....

3. Bi-metal

....
....
....
....
....
....
....
....
....
....

Complete the sectioned views below to show an example of each type.

What is meant by piston slap and when does it most often occur?

....
....
....

INVESTIGATION

1. To determine variation in piston diameter.

   Measure, using a micrometer, a piston at the positions shown below.

2. To determine typical recommended piston/bore clearances.

   Use a workshop manual to determine a piston's cold-working clearance at the positions shown below.

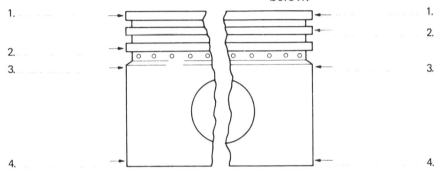

1. ....
2. ....
3. ....
4. ....

1. ....
2. ....
3. ....
4. ....

3. To determine the ovality of typical pistons.

Measure two pistons of different design but of similar size to determine the varying amounts of ovality when cold.

Measure at the positions shown in drawing I.

Type I

Micrometer position	On gudgeon-pin axis	Across thrust faces	Ovality
1.			
2.			
3.			
4.			

Type II

Micrometer position	On gudgeon-pin axis	Across thrust faces	Ovality
1.			
2.			
3.			
4.			

PISTON RINGS

What are the three basic functions of a piston ring?

1. ....
....

2. ....
....

3. ....
....

Petrol engines usually have pistons with two compression rings and one oil-control ring above the gudgeon pin.

Identify the rings shown in the sketch below.

State the number and type of rings commonly fitted on a C.I. engine piston.

....
....
....
....
....

Compression rings

The top ring is usually of a plain rectangular section, with often the outer edge either chromium or molybdenum plated.

The second piston ring may be a plain, taper-faced or internally stepped ring.

Show a section view of each compression ring named below:

(i) Plain chromium plated, (ii) Under cut, (iii) Internally stepped.

Below is shown a section of a large C.I. engine piston. Show on the centre line typical sections of the rings suitable for this type of piston, and name the rings.

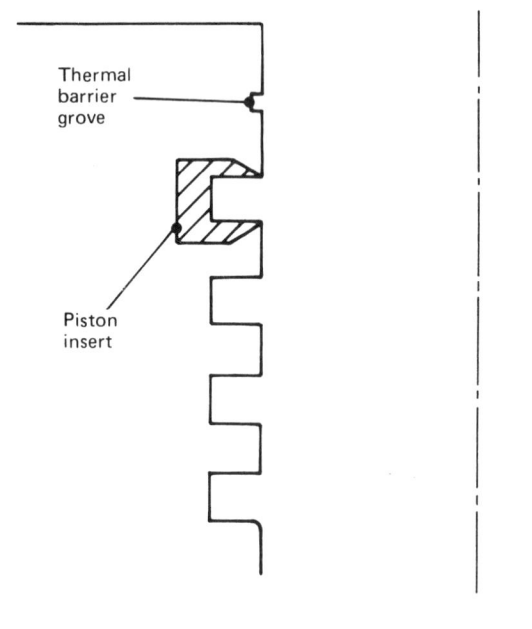

Thermal barrier grove

Piston insert

Why is a piston insert used?

....

....

Oil-control rings

This type of ring should glide over the oil film as the piston moves up the cylinder yet scrape off all but a thin film of oil when descending.

The oil scraper ring above the gudgeon pin on older types of engines is usually of a slotted design.

....

....

....

....

....

....

Draw a sectioned view of a slotted scraper ring in the piston section below.

Oil scraper rings

These are designed to exert a greater radial force on the cylinder wall than the above types, and are usually of a fabricated construction.

Compression rings

Fabricated oil control ring

....

....

....

....

....

....

Draw a section of a **steel rail multi-piece ring.**

36

SPECIAL RINGS USED IN TWO-STROKE C.I. ENGINES

Two unusual types of rings, used mainly on two-stroke C.I. engines, are the wedge compression ring and the fire or junk ring.

Sketch and state the purpose of these rings.

Wedge compression ring ....
....
....
....
....
....

Fire ring ....
....
....
....
....
....

PISTON-RING FAULTS

Explain the major causes of the following faults that can occur to rings in service.

Ring breakage ....
....
....

Ring flutter ....
....
....

Ring click ....
....
....

Gummed rings ....
....
....

Passing too much oil ....
....
....

PISTON FAULTS

Most piston faults can be attributed to piston-ring troubles. There are, however, many other causes of faults:

....
....
....
....
....
....
....
....

INVESTIGATIONS

Piston rings

Measure the individual gaps of a set of rings in their cylinder bore.

Assemble the rings on the piston and measure the ring groove clearance.

Record your findings in the table below.

Engine make .... Model .... Cylinder diameter ....

Ring type and position on piston	Ring gaps		Groove clearance		Serviceability
	Actual	Recommended	Actual	Recommended	

The sketches show where readings should be taken.

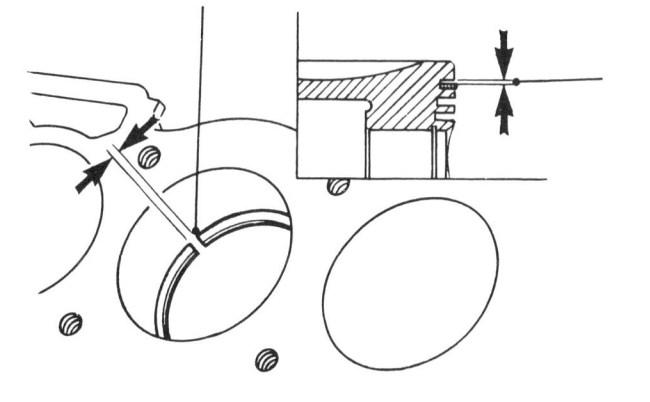

The sketch below shows recommended positions of the ring gaps when fitted on to the piston.

....

....

....

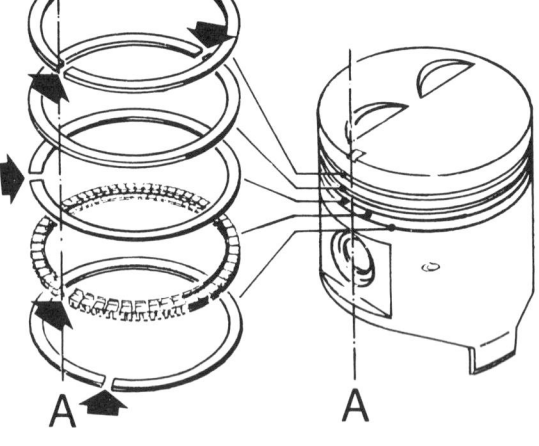

Using a special tool similar to the one shown, fit the piston into the cylinder bore.

State the procedure that should be observed.

....

....

....

....

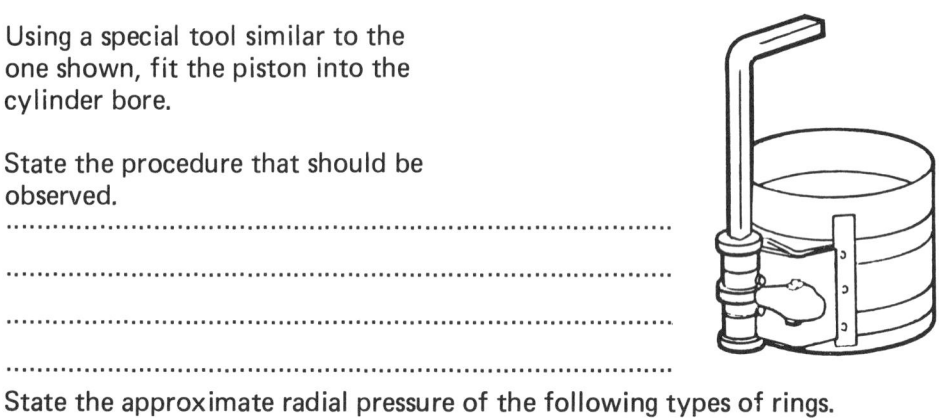

State the approximate radial pressure of the following types of rings.

Compression ....   Normal oil scraper ....

Special oil scraper ....   Internally stepped ....

GUDGEON PINS

The gudgeon pin connects the piston to the connecting rod and has to withstand the shock loads created by the forces of combustion.

The pin must be free to rotate in either the piston or the connecting rod (or both) in order to allow these two components to move relative to one another.

The pin must also be prevented from moving sideways and scoring the cylinder walls. There are four basic ways in which this can be done.

1. Gudgeon pin fully floating held with circlips fitting into the gudgeon-pin bosses.

....

....

....

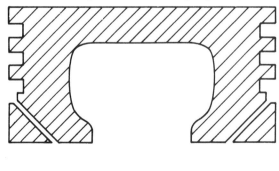

Complete the sectioned diagram to show a gudgeon pin located by a circlip.

 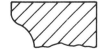

2. A fully floating gudgeon pin may be fitted with end-pads.

Complete the diagram to show a sectioned view of a gudgeon pin fitted with end-pads.
State from what material the end-pad may be made.

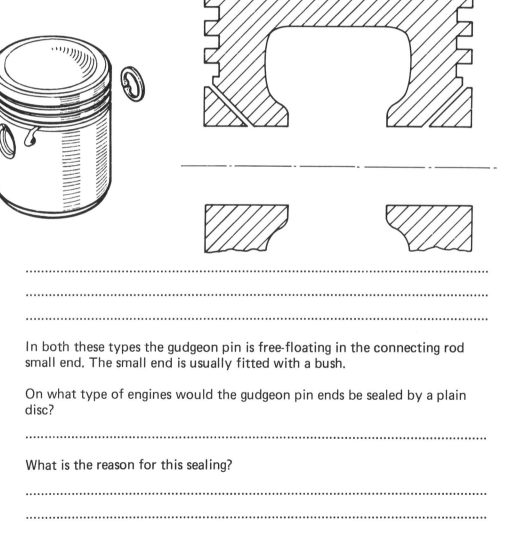

....

....

....

In both these types the gudgeon pin is free-floating in the connecting rod small end. The small end is usually fitted with a bush.

On what type of engines would the gudgeon pin ends be sealed by a plain disc?

....

What is the reason for this sealing?

....

....

3. Gudgeon pin held in small end with a clamp-bolt (pin is a thermal fit in piston).

Complete the drawing below to show a part-sectioned gudgeon pin and clamp-type small end, indicate how oil reaches the piston bosses.

....

....

Show a side view of the small end clamp bolt type

4. Gudgeon-pin interference fit in small end, free in piston.

Complete the drawing below to show a sectioned view of the gudgeon pin having an interference fit in the connecting rod small end. (Note a small end bush is not fitted.)

Why is this type not popular on modern engines?

....

....

....

Why are gudgeon pins normally hollow?

....

....

....

40

CONNECTING-ROD BEARING ARRANGEMENTS

Small end

The small end of the connecting rod locates the gudgeon pin.
In the types 1 and 2 (page 39) the gudgeon pin is fully floating and is fitted with a bush which requires lubrication. Type 3 is a clamp type and type 4 is an interference fit (page 40); in these types the pin does not move in the rod and so no lubrication is required.

The sketches show upper sections of two connecting rods.

Describe how each type is lubricated.

Big end

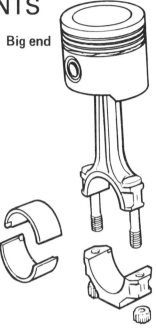
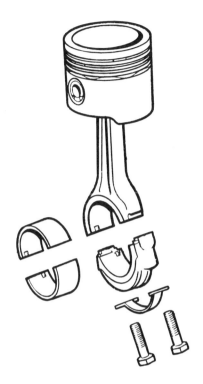

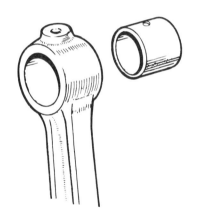

....

....

....

....

....

....

....

Conventional type   **Oblique type**

Identify the various parts of the assemblies shown above.
Describe how the bearings are:

(a) located ....

....

....

....

(b) prevented from rotating ....

....

....

The big end locates on the crankshaft and enables the reciprocating motion of the piston to be transferred into the rotary motion of the crankshaft.

What is the reason for designing the big end so that the cap face split line is at an oblique angle?

....

....

....

What is the disadvantage of splitting the big end at an angle?

....

....

Explain how oil is supplied to the big end bearings and how they supply oil to the cylinder walls.

....

....

....

The rod must withstand the compressive, tensile, twisting and bending forces that are set up as it transfers the to-and-fro motion of the piston into the rotary motion of the crankshaft. Why is the rod section usually of an ⊢ section as shown?

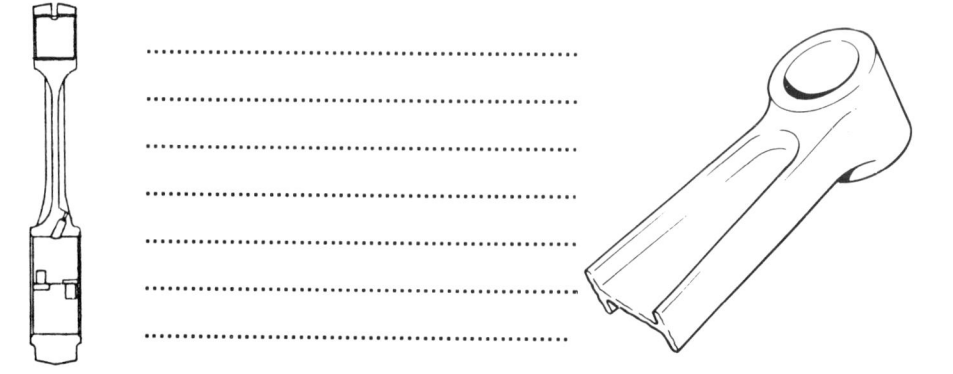

CRANKSHAFTS

State the meaning of the following terms, as applied to crankshafts, and indicate one of each of them on the drawing below.

Main bearing journal ....

....

Crankpin journal ....

....

Throw ....

....

Web ....

....

Journal radius ....

....

Balance weight ....

....

Type ....

FOUR-CYLINDER IN-LINE CRANKSHAFTS

Crankshafts are commonly made from drop forgings.

....
....
....

5-main bearing dropped forged crankshaft.

Complete the diagrams below to show a 5-main bearing crankshaft (the blocks represent the mains bearing journals).

On the end view indicate the crank angle disposition and state which cranks are at t.d.c. and b.d.c.

Note. Number the cylinders from the non-driving end.

An alternative method to produce crankshafts is by casting.

3-main bearing cast crankshaft. The large webs can be hollow.

List the advantages of drop forged crankshafts compared with cast crankshafts.

....
....
....
....

Why is it considered desirable to use 5-main bearings instead of 3?

....
....
....

CYLINDER NUMBERING & FIRING ORDER

There are two practicable firing orders for a four-cylinder in-line engine, these are:

.... and ....

With the aid of these firing orders complete the tables below to show which one of the four strokes in the four-stroke cycle each individual cylinder is on at any given instant.

Table A

Cylinder number	Firing order			
	1		4	
1	P			
2				
3				
4				

Table B

Cylinder number	Firing order			
	1		4	
1	P			
2				
3				
4				

INSTRUCTIONS

1. Complete the firing orders, one in each table.
2. Fill in the table using the code:
   I, induction; C, compression; P, power; E, exhaust.

Give an example of each of the two firing orders above

Make	Model	Firing order

Reading Table A

When 1 is on power stroke 2 will be on ....
When 4 is on exhaust stroke 3 will be on ....

Reading Table B

When 3 is on induction stroke 2 will be on ....
When 1 is on compression stroke 4 will be on ....

FOUR-CYLINDER VEE CRANKSHAFTS

The most common vee-four engine arrangement in Britain uses the 60° vee as shown.

Name a make and model of vehicle using this type of power unit.

Vehicle make ....

Model ....

Show positions of cranks if engine is to be designed to have equal firing intervals.

Number the cylinders below and state a suitable firing order.

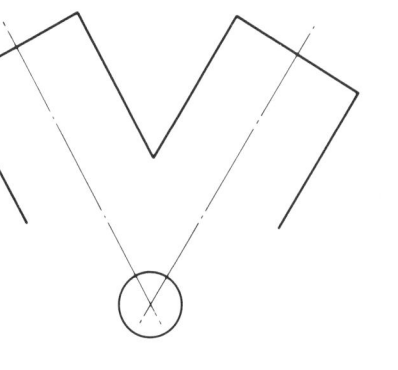

....

What unusual feature does this type of engine require with regard to engine balance?

....
....
....
....
....

SIX-CYLINDER IN-LINE CRANKSHAFTS

The diagrams below show the most common crank throw arrangements.

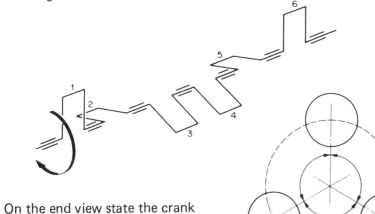

On the end view state the crank angles and number the positions of the crank.

Suitable firing order ....

Complete the line diagram to show an alternatively designed crankshaft. On the end view number the positions of the cranks.

Suitable firing order ....

SIX-CYLINDER VEE CRANKSHAFTS

Show positions of cranks if engine is to be designed to have equal firing intervals.

Number the cylinders below and state a suitable firing order.

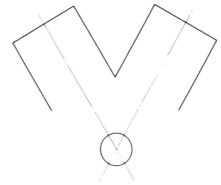

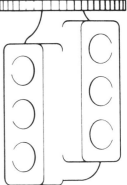

....

....

....

....

....

....

....

....

....

Which in-line firing order is the most popular?

....

....

VEE-EIGHT CRANKSHAFTS

When the vee is a 90° angle and the crankshaft designed as shown, the engine's primary and secondary forces can be perfectly balanced and the firing impulses evenly spaced.

The 90° angle, however, means that the engine is very wide.

If the angle is narrowed to 60° the crank becomes unbalanced and the firing intervals occur at unequal angles.

....
....
....
....
....
....
....

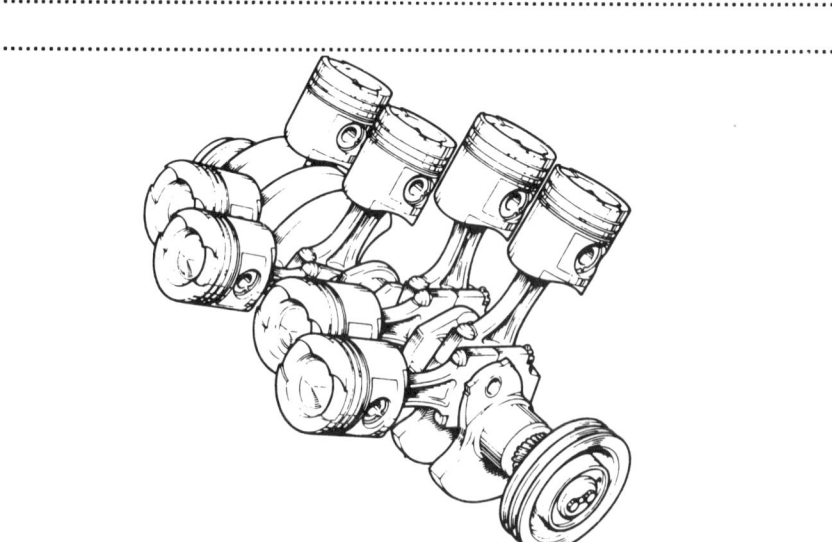

How many main bearings has the V8 C.I. engine shown?

Number the cylinders on the engine block.

Number the crankpins on both views of the crankshaft.

State a suitable firing order.

....

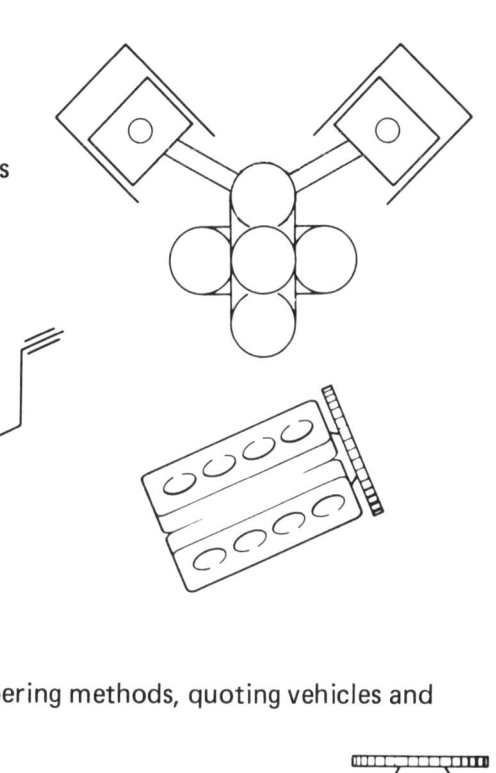

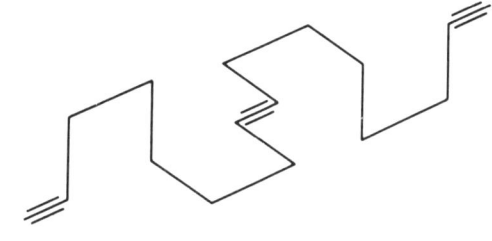

Two-plane crankshaft

Show two alternative cylinder numbering methods, quoting vehicles and firing orders.

Make	Make
....	....
firing order	firing order
....	....

State the advantages of using vee engines compared with in-line types.

....
....
....
....
....
....

CHECKING CYLINDER BORE WEAR

The diameter of cylinder bores may be checked by ....

An internal micrometer takes direct readings but requires a sensitive touch to obtain accurate readings. When a cylinder bore gauge is used it must be calibrated by using either a ring gauge or an external micrometer.

The cylinder bore gauge converts the horizontal movement of the spring-loaded plunger into a vertical movement, which is transferred by a push rod (in the gauge shaft handle) to a dial test indicator clamped to the top of the handle.

Describe how the cylinder bore wear may be checked, using such a gauge.

....
....
....
....

Having obtained the original bore diameter and maximum wear, state two other values that can be measured.

....
....
....

How are larger bores checked, using the same gauge?

....
....

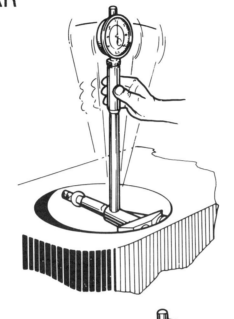

Setting cylinder bore gauge to read

....

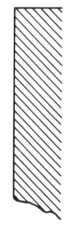

Sketch an internal micrometer, taking a reading in the maximum wear region of the bore.

Name the main parts.

Indicate by shading the normal bore wear.

ATTACHMENT OF COMPONENTS TO THE CRANKSHAFT

The timing gear and crankshaft pulley are usually fitted at the front end of the crankshaft.

Examine such an arrangement and complete the sketch below to show their location, how they are secured, and also show the timing cover oil seal.

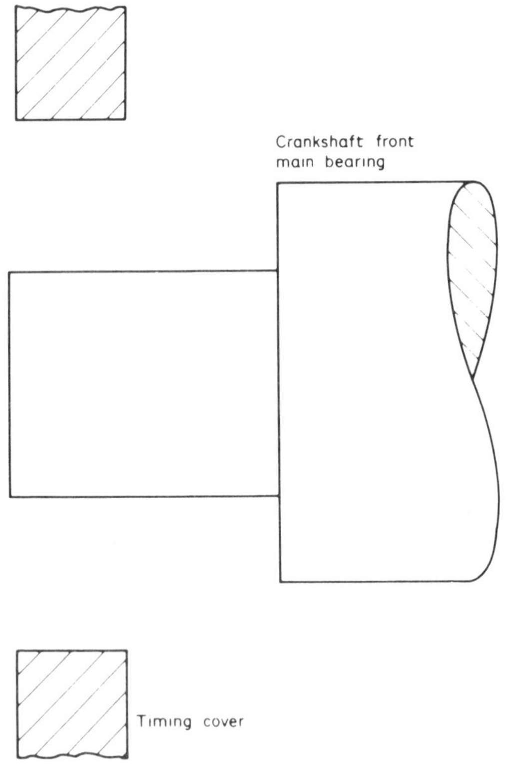

Crankshaft front
main bearing

Timing cover

What oil sealing method, alternative to the one shown above, may be used?

....

....

....

Examine various engines and show two alternative methods of attaching the flywheel to the crankshaft.

Show clearly any dowel keyway or locking washer arrangement and the method of oil sealing where possible.

Engine make ....

Model ....

Type of fixing

....

Engine make ....

Model ....

Type of fixing

....

What is the object of fitting dowels as well as studs to the flange type flywheel mounting?

....

....

....

48

CRANKSHAFT TORSIONAL VIBRATION

It is easy to realise that metal components vibrate with a certain natural frequency. Many factors contribute to the precise vibratory frequency, one of which is length.

This can easily be demonstrated by clamping one end of a short steel rule and flicking the free end.

Do the same with a longer steel rule and note the result.

The short steel rule had a .... natural vibratory frequency.

The long steel rule had a .... natural vibratory frequency.

State the causes of torsional vibration set up in a crankshaft.

....

....

....

....

....

....

....

Why is it important to stop vibrations from building up, even in a massive component such as a crankshaft?

....

....

....

....

CRANKSHAFT VIBRATION DAMPERS

The following is a list of different types of engines. State which you consider would have (a) the greatest and (b) the least need of a crankshaft vibration damper and why this is so.

In-line four-cylinder; vee-six; in-line six-cylinder; in-line two-cylinder.

(a) Engine most likely to need a crankshaft vibration damper.......................
   because

....

....

....

(b) Engine least likely to need a crankshaft damper ....
   because

....

There are three basic types of vibration dampers:

1. .... 2. .... 3. ....

PRINCIPLE OF OPERATION – RUBBER TYPE

....

....

....

....

....

....

....

Make a sectional sketch of a crankshaft vibration damper fitted on either a car or commercial vehicle, use the centre lines provided as a guide and name the main parts.

Vehicle make .... Model ....

Type of damper ....

The graph below shows the amount of oscillation build-up on a particular six-cylinder engine when the natural frequency of the crankshaft is at the same frequency as one of the 'critical' impulses of combustion.

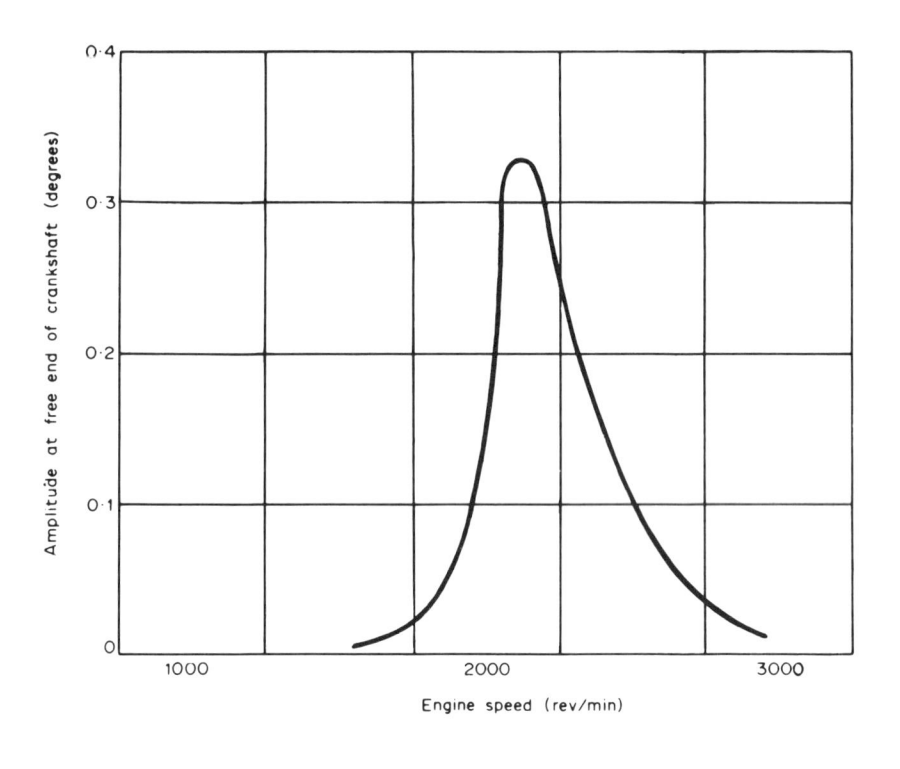

Show on the graph, the likely affect that the fitting of a vibration damper would have on the amplitude of vibration.

....
....
....
....
....
....

CRANKSHAFT ALIGNMENT

After a period in service a crankshaft will become worn on its journals. This can be recognised and measured relatively easily.

Maximum wear is ....

....

Two other types of wear are commonly found on crankshaft journals. On each sketch below, indicate the shape (exaggerated) of one type of wear, name it and show where measurements would be taken.

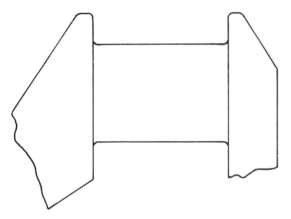
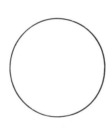

The crankshaft may also suffer from a less obvious defect such as becoming out of alignment, due to twist and/or bend (or bowl).

Title appropriately the two diagrams below:

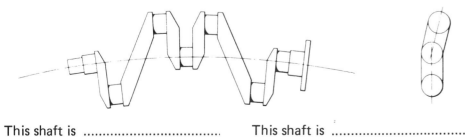

This shaft is ....   This shaft is ....

PRINCIPLES OF CHECKING CRANKSHAFT ALIGNMENT

Set up a crankshaft between vee blocks as shown and check for bend and twist.

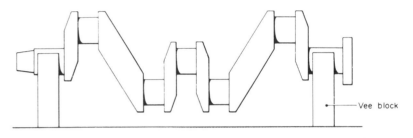

Vee block

With the aid of the above diagram, explain how the shaft would be checked for bend.

....

....

....

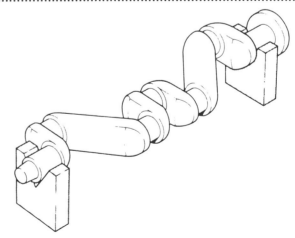

Indicate on the above drawing where readings should be taken to check the shaft for twist.

ENGINE MOUNTINGS

The engine/transmission unit is attached to the chassis or body structure through rubber blocks. Their main function being to:

....

Opposite is shown a conventional engine/transmission layout.

Examine such a layout and then examine a transverse engine mounting layout such as a Mini.

Describe any differences in their basic mounting arrangements.

....
....
....

In what way does the conventional commercial vehicle engine mounting arrangement differ from the one shown opposite?

....
....
....

The mountings are considered to be fail safe.

What is meant by this expression in this case?

....
....
....

What are the *basic* causes of engine vibration?

....
....
....

The sketches below illustrate a typical 'three-point' mounting arrangement for an engine and gearbox assembly of conventional layout.

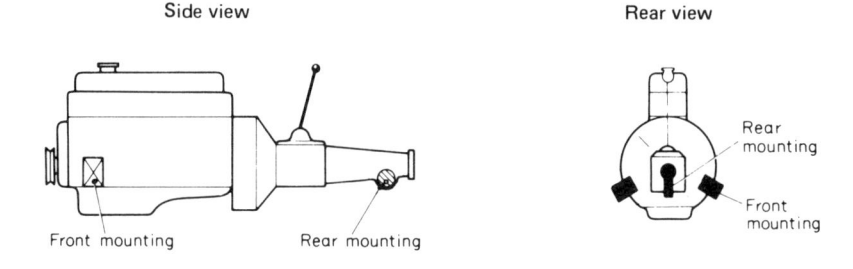

Side view   Rear view

Front mounting   Rear mounting

Rear mounting

Front mounting

Below are shown two typical mountings as may be used at B.
In each case indicate the mounting and its metal/rubber component.

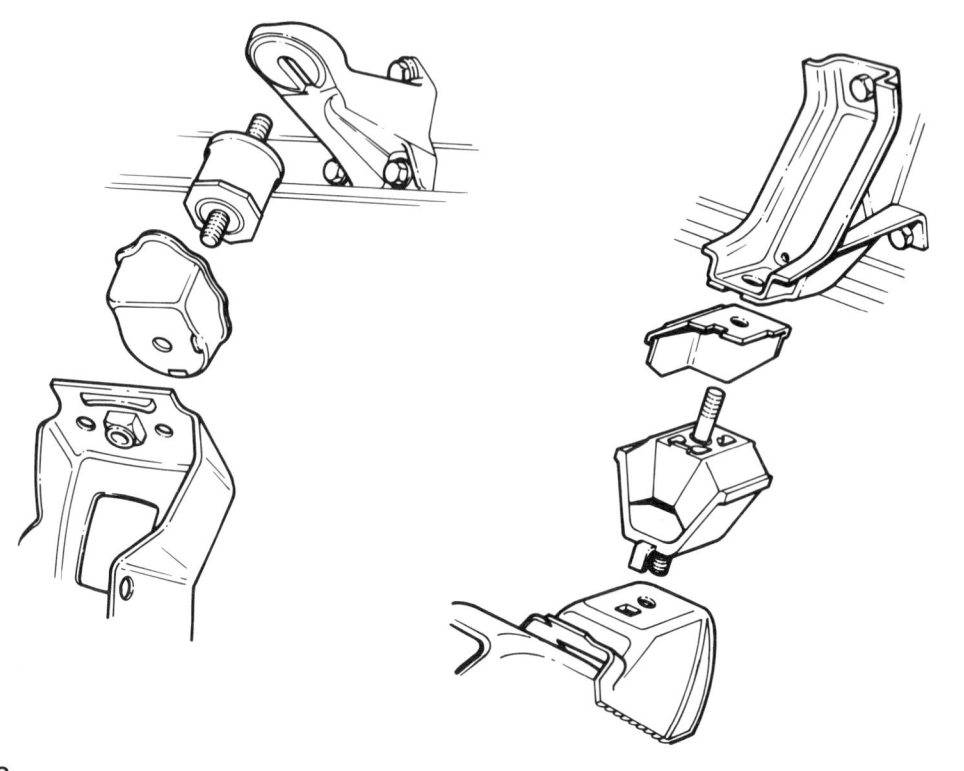

ENGINES – APPLIED STUDIES

END-FLOAT – CAMSHAFT

In order to revolve freely at speed the camshaft must possess an end-float clearance. Typical limits are: ....

The clearance may be checked, as shown on the overhead camshaft arrangement below, using feeler gauges or by using

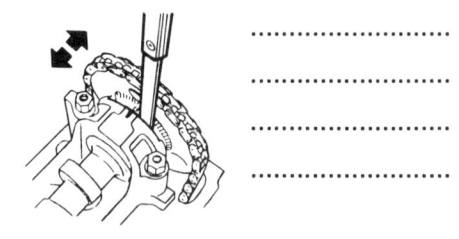

....

....

....

....

The correct camshaft end float on four similar engines is 0.05 mm, Complete the table to show the extra thickness of pad required to bring the end float to the recommended value.

Measured end-float	0.27	0.58	0.34	0.05
Extra thickness required				

END-FLOAT – CRANKSHAFT

Crankshaft end-float should normally be between limits of ....

The clearance is checked in a similar manner to the camshaft.

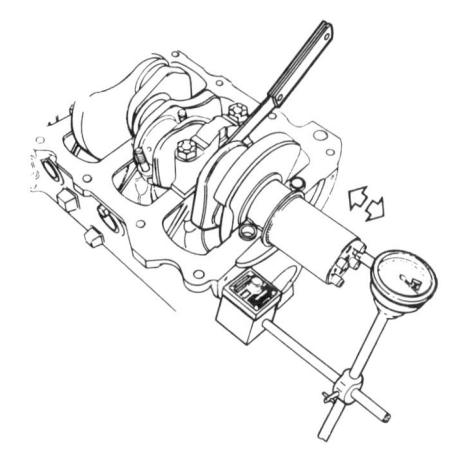

How can the end-thrust be altered?

....

....

The end-float clearance when measured as shown was 0.45 mm but the correct end float should be 0.15 mm. The thickness of the thrust washers when removed was found to be 2.85 mm each. What should be the thickness of the new thrust washers?

CRANKSHAFT-TO-CAMSHAFT MOVEMENT RATIO

The crankshaft-to-camshaft movement ratio in all four-stroke engines is ....

On two-stroke C.I. engines using exhaust valves the movement ratio is ....

On simple camshaft drive arrangements this ratio can be proved mathematically by either counting the number of teeth on each gear or by measuring the gears diameters and dividing the driver gear value by the driven gear value.

Examine an engine with a drive arrangement similar to those shown below.

Calculate their movement ratios by:

1. Counting the number of teeth on each gear.

2. Measuring the diameter of each gear wheel.

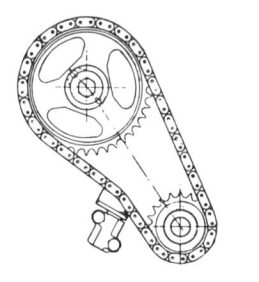

On certain vehicles a twin-chain arrangement is used and the gear sizes do not need to be in a strict double size ratio.

If the camshaft gears on the diagram shown have 30 T and intermediate gears have 28 T and 20 T calculate the number of teeth on the crankshaft gear.

Camshaft gears

Intermediate gears

Crankshaft gears

FORCES ACTING ON VALVE MECHANISMS

Engineering materials are designed to withstand the external forces and pressures placed upon them. The strength of a material is determined by its ability to resist loads without breaking.

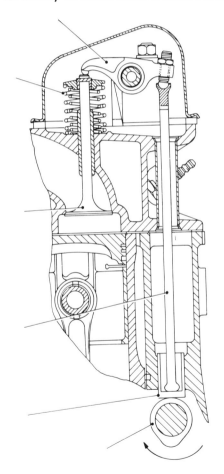

Identify the types of forces created in the valve mechanism shown by the rotating camshaft and coil return springs.

....
....
....
....
....
....
....
....
....
....
....
....
....
....
....
....
....
....

Name the main types of force which act on the indicated items.

STRESS

The internal reaction that is set up in a material when force is applied is known as STRESS and the (often small) deformation that takes place when stress occurs is known as STRAIN.

State the three basic types of stress.

....

State the principle type of stress to which the following engine components are mainly subjected:

Component	Type of stress
Cylinder-head bolt	
Timing chain	
Camshaft	
Crankshaft	
Connecting rod	
Gudgeon pin	

The stress induced in a material is directly proportional to the force applied. It is usual to express the stress in terms of force per unit area.

Stress =

PROBLEM

Determine the compressive stress induced in a push rod of 30 mm cross sectional area when a force of 150 N is applied to move the rod.

Stress =

MATERIALS USED IN ENGINE CONSTRUCTION

State the materials used in the components shown and in each case give a reason for their choice.

Valve spring ....

....

....

Valve guide ....

....

....

....

Valves ....

....

....

Cylinder block ....

....

....

....

Crankshaft ....

....

....

Timing chain tensioner ....

....

....

Rocker cover ....

....

....

....

....

Rocker cover gasket ....

....

....

Cylinder head ....

....

....

....

Cylinder liner ....

....

....

Gudgeon pin ....

....

....

....

Connecting rod ....

....

....

CYLINDER HEAD GASKETS

Name the materials used and describe different types of cylinder head gaskets.

....

....

....

....

....

....

....

....

....

....

....

....

What are the basic causes of head and cylinder block mating face distortion?

....

....

....

....

....

PISTONS

State the most common material used in the manufacture of pistons; give reasons for this choice.

....

....

Other materials used to a much lesser extent are:

Aluminium copper alloy (Y alloys) ....

Cast iron ....

Steel alloy ....

Pistons using more than one material. ....

....

PISTON RINGS

Give reasons for the choice of material used in:

Compression rings ....

....

....

....

....

Oil control rings ....

....

....

....

CAMSHAFT BEARINGS

When the camshaft is mounted in a cast-iron crankcase the bearings may be formed simply by borings in the casting itself; or more commonly by ....

....

What is the most common form of bearing arrangement on overhead camshaft engines?

....

....

....

....

MAIN AND BIG-END BEARINGS

Modern main and big-end bearings are almost always of the precision-fit, thin-wall shell type. What features make them so popular?

....

....

....

Examine a thin-walled metal bearing. State the layers of material from which it is made. Give the approximate thickness of each layer and say what is its function.

Material	Function	Approximate thickness

The lining material of the bearing varies depending on the loads it must withstand.

State the bearing materials used, for a typical application on the following types of engines:

Engine type	Bearing lining material
Low-powered petrol	
High-powered petrol	
Compression-ignition	

....

....

....

....

....

State some of the factors which cause bearing damage on failure:

....

....

....

....

....

....

....

THERMAL EXPANSION

When materials, solid or gas, are heated they expand and enlarge their volume.

List the components in an engine which must be given clearance to allow for thermal expansion.

....

....

....

Certain components obtain a clearance when heated because the mating parts are made from different metals.

....

....

....

The reason for this is that different materials have different rates of expansion. This is known as the

....

Give the linear expansion rate for:

Aluminium ....

Brass ....

Steel ....

Cast iron ....

In some instances advantage is taken of expansion or contraction when assembling engine components; give some examples.

....

....

....

LIMITS AND FITS

It is essential when manufacturing motor vehicle components that they are accurately made, readily interchangeable, and capable of rapid assembly.

The type of fit can be stated, when, within the limits of the shaft and hole size, every possible fit has been considered.

Define the following fits and show diagrammatically a shaft of appropriate size to fit the holes. Show the tolerance and indicate the maximum and minimum fits.

Give motor vehicle examples of each fit.

CLEARANCE FIT

....

....

Examples ....

....

INTERFERENCE FIT

....

....

Examples ....

....

TRANSITION FIT

....

....

Examples ....

....

59

CLEARANCE FITS

Examine the drawing you produced to show a clearance fit. This should show that:

Maximum clearance = largest hole — smallest shaft

Minimum clearance =   hole —   shaft

Examples

1. A main bearing shell is machined to $50 \cdot 50 \begin{array}{c} + 0 \cdot 03 \\ - 0 \cdot 01 \end{array}$ mm diameter and the crankshaft journal on which this bearing has to fit is machined to $50 \cdot 50 \begin{array}{c} - 0 \cdot 05 \\ - 0 \cdot 07 \end{array}$ mm diameter. Calculate the maximum and minimum clearance.

....

....

....

....

....

2. Find the maximum and minimum clearance between a crankshaft journal $60 \pm 0 \cdot 02$ mm diameter and bearing of $60 \begin{array}{c} + 0 \cdot 09 \\ + 0 \cdot 05 \end{array}$ mm diameter.

....

....

....

....

....

INTERFERENCE FITS

Examine the drawing you produced to show an interference fit. This should show that:

Maximum interference = largest shaft —..................hole.

Minimum interference = ... shaft — .... hole.

Examples

1. A cylinder block is bored out to $95 \begin{array}{c} + 0 \cdot 03 \\ - 0 \cdot 01 \end{array}$ mm diameter and a cylinder liner is machined to $95 \begin{array}{c} + 0 \cdot 20 \\ + 0 \cdot 15 \end{array}$ mm diameter. Calculate the maximum interference fits.

....

....

....

....

....

....

....

2. Determine the maximum and minimum variation of fit between a cylinder liner $100 \begin{array}{c} + 0 \cdot 15 \\ + 0 \cdot 10 \end{array}$ mm diameter and a $100 \begin{array}{c} + 0 \cdot 03 \\ - 0 \cdot 00 \end{array}$ mm cylinder block diameter.

....

....

....

....

....

....

TAPER FITS

The crankshaft timing gear and pulley are commonly located by a woodruff key on a parallel shaft.

When is it necessary to use a taper to secure timing gears or pulleys?

....

....

....

....

....

60

VALVE TIMING DIAGRAMS

A valve timing diagram shows in degrees of crank angle the total period in the engine cycle during which each valve is open and the points at which it opens and closes.

With the aid of a protractor measure the valve open periods and complete the tables below.

I.V.O. ....

I.V.C. ....

Total inlet valve open period

....

E.V.O. ....

E.V.C. ....

Total exhaust valve open period

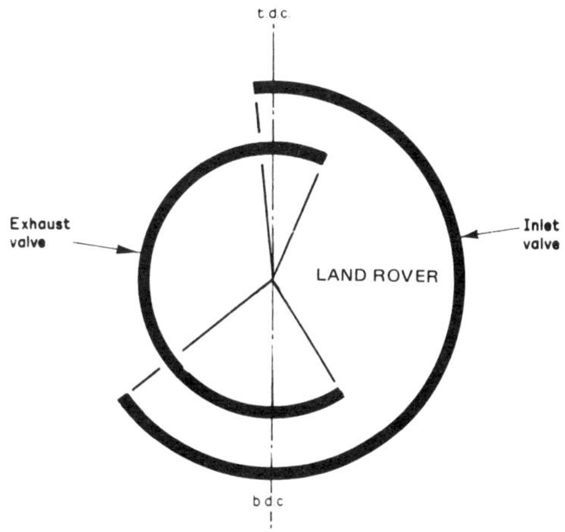

....

I.V.O. ....

I.V.C. ....

Total inlet valve open period

....

E.V.O. ....

E.V.C. ....

Total exhaust valve open period

....

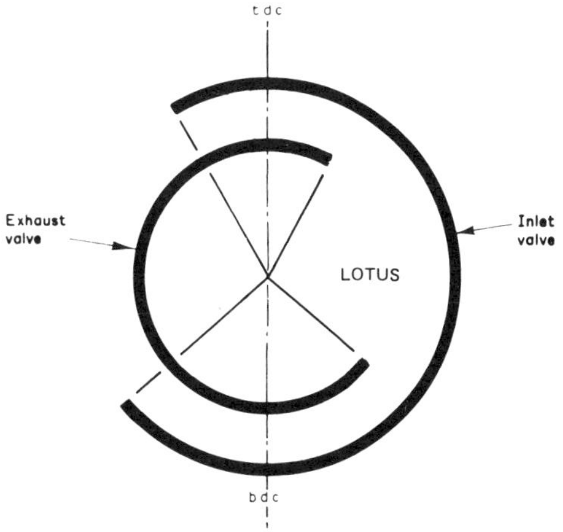

Examine the valve timing specifications for similar capacity spark-ignition engines in different stages of engine tune.

(Most major manufacturers produce a range of such engines.)

Write in the specifications obtained and construct and label valve timing diagrams on the centre lines below.

Engine

Make ....

Model ....

I.V.O. ....

I.V.C. ....

Total inlet valve open period

....

E.V.O. ....

E.V.C. ....

Total exhaust valve open period

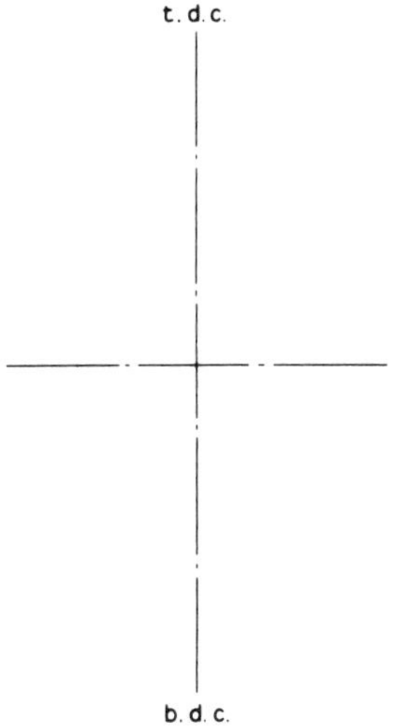

....

Engine

Make ....

Model ....

I.V.O. ....

I.V.C. ....

Total inlet valve open period

....

E.V.O. ....

E.V.C. ....

Total exhaust valve open period

....

61

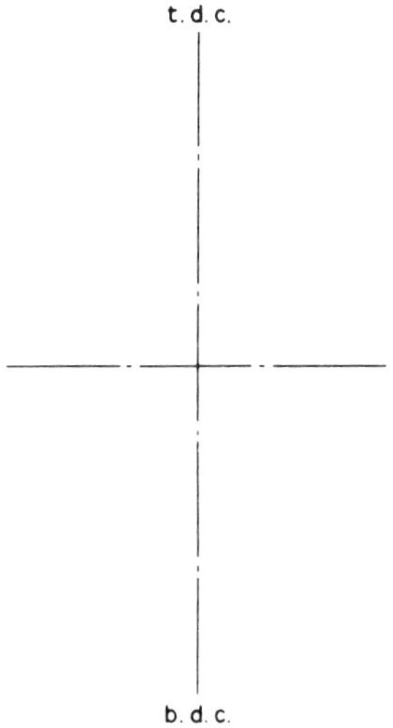

COMPRESSION-IGNITION, FOUR-STROKE CYCLE

The valve timing is similar to the spark-ignition engine timing, but since the fuel is injected over a specific period it too should be shown.

Obtain the valve timing specification of a C.I. engine and draw the valve timing diagram.

Vehicle make .... Model ....

Engine type .... Capactity ....

Valve timing specifications

I.V.O. .... b.t.d.c.   E.V.O. .... b.b.d.c.

I.V.C. .... a.b.d.c.   E.V.C. .... a.t.d.c.

Spill point .... b.t.d.c.

Total inlet valve
open period = ...

Total exhaust valve
open period = ...

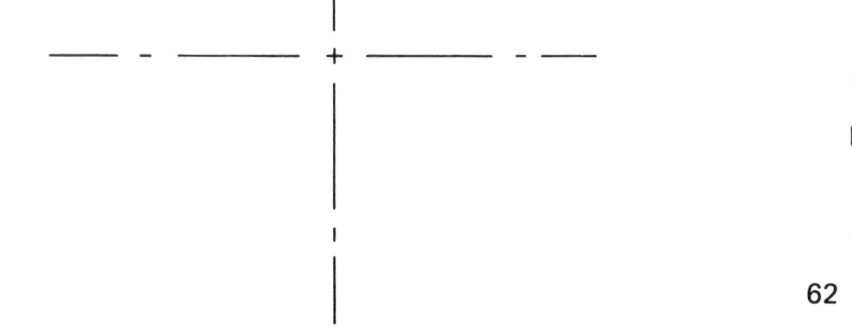

VALVE-PORT TIMING DIAGRAMS TWO-STROKE SPARK-IGNITION

All the operations occur in one crankshaft revolution.

Engines using ports only and employing the crankcase as an induction pressure chamber always have equal port openings either side of t.d.c. or b.d.c.

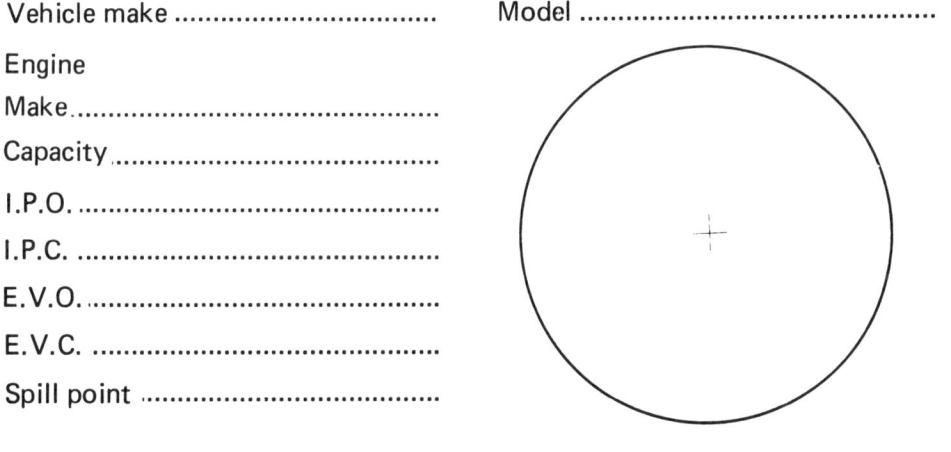

Typical two-stroke timing diagram

TWO-STROKE COMPRESSION-IGNITION

These have a similar diagram but without the transfer port at the top.

Obtain the valve timing specifications for a two-stroke C.I. engine using INLET PORTS AND EXHAUST VALVES and complete the diagram and table below.

Vehicle make .... Model ....

Engine
Make...

Capacity...

I.P.O. ....

I.P.C. ....

E.V.O. ....

E.V.C. ....

Spill point ....

62

VALVE TIMING CALCULATIONS

If the valve-open period of an engine is required and no data are given, it will be necessary to measure the open period distance on the rim of the flywheel and convert this distance to degrees.

Length of arc
(no. of mm)

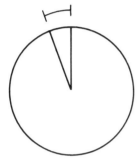

Angle of arc
(no. of degrees)

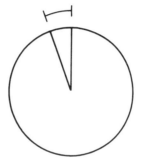

Circumference
(πd)

Total angle
of circle 360°

Consider the ratios above

$$\frac{\text{length of arc}}{\text{circumference}} \qquad \text{and} \qquad \frac{\text{angle of arc}}{\text{total angle of circle}}$$

i.e. $\dfrac{\text{no. of mm}}{\pi d}$ and $\dfrac{\text{no. of degrees}}{360}$

These ratios are proportional to one another and may be written as:

Using this formula
   the length of arc around the flywheel can be determined

length of arc =
   (no. of mm)

or
   the angle of the arc can be determined

angle of arc =
(no. of degrees)

Problems

1. The length of arc from t.d.c. to the point where the inlet valve just opened was 18 mm measured on the outer edge of the flywheel. Convert this distance to degrees.

   The flywheel diameter being 270 mm.

2. The length of arc from b.d.c. to the point where the inlet valve just closed was 108 mm measured on the outer edge of the flywheel. Convert this distance to degrees.

   The flywheel diameter being 270 mm.

3. An exhaust valve timing data gave the following values:

   opens 54° b.b.d.c, closes 12° a.t.d.c.

   Convert these values into distances on a flywheel rim whose diameter is 280 mm.

4. A valve on an engine closes 225 mm a.b.d.c. and opens 50 mm b.t.d.c. measured on the flywheel circumference whose diameter is 400 mm.

Calculate the number of degrees through which the valve is open.

6. If a valve opens 6° b.t.d.c. and closes 40° a.b.d.c., what is the valve-open period measured in mm around the circumference of the flywheel if the diameter is 350 mm?

8. An inlet valve opens 8° b.t.d.c. and closes 45° a.b.d.c. The exhaust valve opens 45° b.b.d.c. and closes 10° a.t.d.c.

(a) calculate the valve overlap in (i) degrees (ii) mm measured on the flywheel circumference when the diameter is 315 mm.

(b) from the information given, construct a a circular type valve timing diagram.

7. A crankshaft pulley of 175 mm diameter has four timing marks scribed on its circumference. There is an equal angular arc of 6° between each mark.

If the first mark indicated t.d.c. calculate the distance measured on the pulley's circumference to the third mark.

5. A crankshaft pulley of 210 mm diameter, has a timing mark scribed on its circumference 16.5 mm b.t.d.c.

Convert this measurement to degrees.

VALVE MOVEMENT IN RELATION TO CAMSHAFT ROTATION
INVESTIGATION

Use an engine on a suitable stand.

Specified tappet settings. ....Inlet ....Exhaust ....

1. Adjust the clearance on say the exhaust valve to the manufacturer's recommended setting.

2. Fit a degree disc to the flywheel with zero in line with t.d.c.

3. Turn the engine until the piston is at t.d.c. with both valves closed.

4. Mount a dial gauge on the cylinder head, with the plunger in contact with the valve spring collar and zero the gauge.

5. Turn the engine in the normal direction of rotation until the valve begins to open, as shown by the gauge needle. Turn engine back 20°.

6. Rotate the engine and record the valve lift for each 30° of crankshaft rotation until the valve closes.

7. Complete the table of results opposite and draw a graph of valve movement against camshaft rotation.
   (N.B. camshaft rotation is half crankshaft rotation.)

8. Use the dial gauge to determine the opening and closing points of the valve in relation to t.d.c. and record these below.

9. Adjust the valve clearance to double that recommended and again check the opening and closing points of the valve; record these below and indicate their position on the graph.

valve clearance		
Ex. valve opens	.... b.b.d.c.	....................b.b.d.c.
Ex. valve closes	.... a.t.d.c.	.... a.t.d.c.

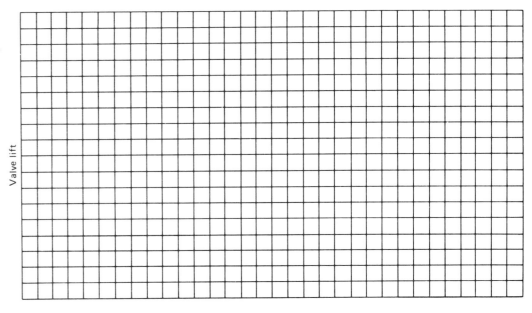

Camshaft rotation

Valve lift

Crank angle											
Cam angle											
Valve movement											

How is the valve open period affected if the valve clearance is incorrect?

(i) Excessive clearance ....
   the valve open period

(ii) Insufficient clearance ....
   the valve open period

How is the maximum valve lift affected by excessive valve clearance?

....

What angular movement would the crankshaft turn in relation to the above graph?

....

PISTON-CRANK MOVEMENT RELATIONSHIP

Show the distance moved during each quarter stroke of the piston.

Comment on how the linear speed varies during half the cranks rotation.

....

....

....

When the crank has moved 90° past t.d.c., the piston will have moved

....

....

....

DETERMINATION OF TOP DEAD CENTRE

Determine, using a suitable engine, how the position may be obtained when no timing marks are available.

Describe the procedure carried out on the engine.

....

....

....

....

....

....

....

....

TORQUE TRANSMITTED TO CRANKSHAFT

The force of combustion pushes the piston down its stroke. If the average force is known when the connecting rod and crank are at right angles, the torque transmitted to the crankshaft can easily be calculated.

Calculate the torque (Nm) applied to the crankshaft in each situation shown.

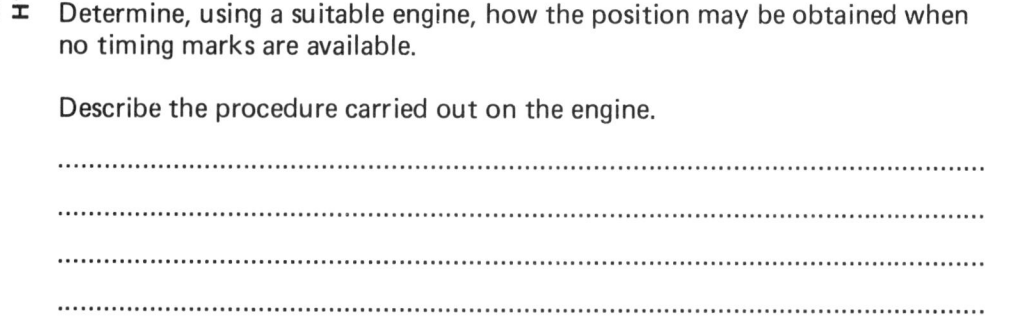

$F = 2500$ N $F = 3200$ N $F = 4.5$ kN

$r = 0.04$ m $r = 0.036$ m $r = 55$ mm

....

....

66

COMPRESSION RATIOS

One of the most important design factors of an engine is the compression ratio.

Show typical compression ratio values in the table below.

Spark-ignition			Compression-ignition	
Family saloon car	Sports car	Racing car	Direct injection	Indirect injection

The compression ratio is the number of times the total volume of air, when the piston is at b.d.c. can be compressed into the combustion chamber space by the piston as it reaches t.d.c.

What would be the compression ratio of a cylinder having a clearance volume of the size shown below.

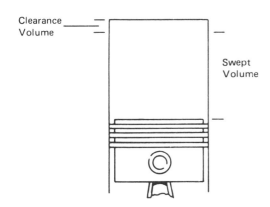

Clearance Volume

Swept Volume

$$\text{Compression ratio} = \frac{\text{swept volume} + \text{clearance volume}}{\text{clearance volume}} = \qquad =$$

Problems

1. An engine has a bore of 80 mm and a stroke of 77 mm. If the volume of the combustion chamber (Vc) is 60 cm^3 determine the compression ratio.

2. Calculate the C.R. of an engine having a clearance volume of 30 cm^3 and a bore and stroke of 70 mm and 60 mm respectively.

3. The bore and stroke of an engine are 84 mm and 100 mm respectively. Calculate the C.R. if the combustion chamber volume is 60000 mm^3.

4. An engine has a bore of 90 mm and a stroke of 140 mm. If the C.R. is 7:1 calculate in cm^3 the clearance volume.

5. Calculate the length of stroke of an engine having a C.R. of 16:1 clearance volume of 72 cm³ and a bore of 10 cm.

6. A four-cylinder engine has a bore of 60 mm and a stroke of 98 mm. What is the total capacity of the engine and the volume of each combustion chamber to give a 9 to 1 CR?

INVESTIGATION

Several methods can be used to find the C.R. of an engine in a practical manner.

Conduct such an experiment and sketch the equipment used in the space below.

Engine make .... Model ....

Length of stroke	Bore diameter	Volume of cylinder	Clearance volume	Compression ratio	Maker's figure

....

....

....

....

....

....

....

SUPERCHARGING

On a normally aspirated engine the volume of a gas entering a cylinder is less than the cylinder volume. The ratio of the volume of gas that could be taken in, compared to the volume that actually enters the cylinder is known as the 'volumetric efficiency' of the engine. It is possible to improve the volumetric efficiency of an engine, and thus the power, by installing a supercharger.

The graph below shows typical engine performance curves for a non-supercharged petrol engine. Show how the curves could be expected to alter if the engine was supercharged.

....

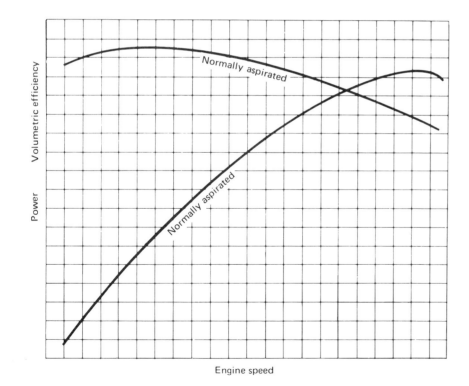

What are the basic differences between supercharging a petrol engine and a compression-ignition engine?

....

....

....

....

....

State a typical boost pressure. ....

....

....

....

EFFECTS OF COMPRESSION RATIO

Explain why it may be necessary to alter the compression ratio of an engine when installing a supercharger.

....

....

....

....

If only a small but still significant increase in power is required, without the necessity, complication or cost of fitting a supercharger, what would be one of the first stages of tune?

....

....

....

....

....

69

ENERGY CONVERSION

The fuel used in a vehicle contains potential heat energy which is stored in a chemical form.

When liberated this heat energy can readily be converted into a mechanical force to propel the vehicle.

The SI unit for heat energy is the ....

Different fuels release differing amounts of heat energy. The calorific value of a fuel (or energy of combustion) is the quantity of J or kJ available to do work when 1 kg of fuel is properly burned.

Complete the following calorific value table.

Fuel	Calorific value
Petrol	
Diesel fuel	
Alcohol	
Paraffin	
Benzol	

Explain why in an engine it is not possible to convert all the heat into useful work?

....

....

....

The percentage of heat used to do useful work from the fuel used in an engine is approximately .... spark-ignition and .... compression-ignition.

This percentage is known as the ....

Heat liberated = mass of fuel x calorific value
$$\text{kg} \quad \text{x kJ/kg}$$
$$= \text{kJ}$$

since 1 kJ = 1 Nm the heat energy can be expressed in terms of mechanical work.

In terms of power the amount of heat energy produced is the amount of heat produced per second.

Energy = fuel consumption (   ) x C.V. (   )

$$= \qquad\qquad =$$

1(a). Calculate the amount of heat energy liberated when 20 kg of fuel are burned. The calorific value of the fuel is 45 000 kJ per kg.
Heat liberated = mass of fuel x calorific value of fuel.

1(b). If it takes 1 h to consume this amount of fuel, what power is available during the process?

2. A certain bearing in a petrol engine absorbs 1500 watts. Calculate the heat energy per min wasted by these bearings.

3. Calculate the quantity of heat energy liberated when 12 kg of fuel is consumed by an oil engine. One kg of oil when burned gives out 40 250 kJ. What power does this represent if the fuel is completely burnt in 30 min.

4. The fuel consumption of an engine is 0.008 kg/s and the C.V. of the fuel is 45 000 kJ/kg. Calculate the brake power if 25% of this energy is converted into useful work.

70

OCTANE RATING

The octane number of a fuel is a measure of the fuel's resistance to knock or detonate under pressure.

Explain how the octane rating of a fuel influences detonation.

....

....

....

....

The octane rating of a fuel is determined by comparing the fuel's resistance to knock with a mixture of *iso*-octane and heptane, both being used in a special engine which is able to vary its compression ratio.

Describe this process ....

....

....

....

....

....

A chemical method is used when values higher than 100 are required.

British Standard 4040:1978 has developed a star classification system which gives an indication of the octane number of a fuel.

State the values which the star ratings indicate.

Star indication	**	***	****
Octane rating	....	....	....

ADDITIVES IN PETROL

The basic elements of all fuels are the elements .... and .... When these are united with oxygen normally taken from the air and ignited heat is produced. Hence the term 'heat engines'. The liquid hydrocarbon fuels used for motor-vehicle engines are produced by distillation from crude oil. One such fuel is petrol, which has a composition of about ....% of ....and .... % of ....

To improve the qualities of petrol when it has been distilled from the crude oil various chemicals are added to it. These are known as fuel additives. One well-known additive is tetra-ethyl lead.

What is its basic purpose in the fuel?

....

....

....

It is now used in smaller quantities than in the past because of its harmful exhaust-emission properties.

State the purpose of other major fuel addititives.

....

....

....

....

....

....

....

71

AIR:FUEL RATIO AND IGNITION SETTING

State the effects of varying the air:fuel ratio and ignition setting on the items below:

Condition \ Item		Power output	Fuel Consumption	Flame speed	Engine temperature	Component life
Air: fuel ratio	Rich					
	Weak					
Ignition Setting	Advanced					
	Retarded					

COMBUSTION AND ENGINE PERFORMANCE

State the effects on combustion and engine performance by varying the items below:

Condition \ Item	Compression ratio	Air:fuel ratio	Ignition timing	Degree of turbulence	Quantity of residual exhaust gas
Combustion					
Engine performance					

COMBUSTION SEQUENCES

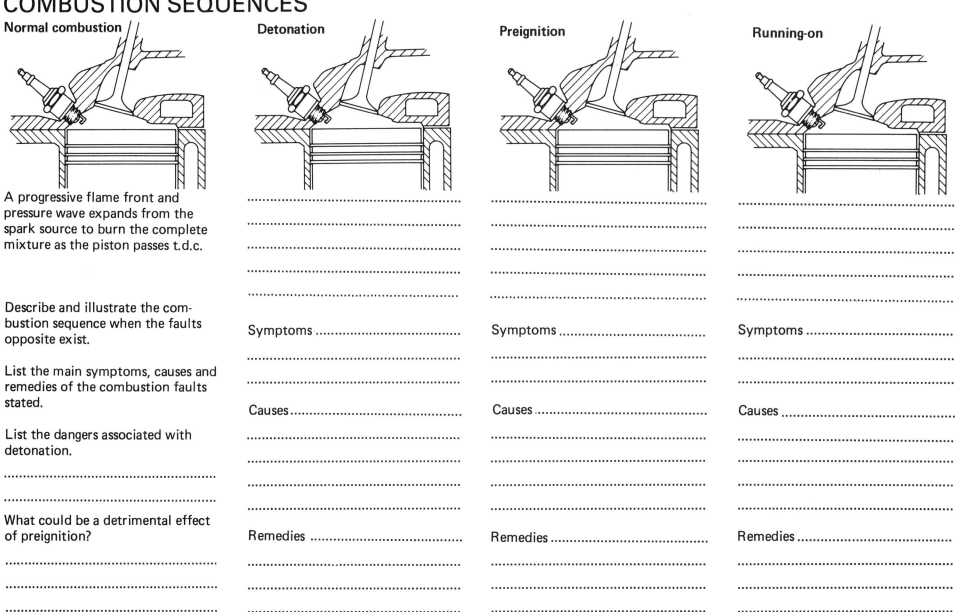

Normal combustion

Detonation

Preignition

Running-on

A progressive flame front and pressure wave expands from the spark source to burn the complete mixture as the piston passes t.d.c.

Describe and illustrate the combustion sequence when the faults opposite exist.

List the main symptoms, causes and remedies of the combustion faults stated.

List the dangers associated with detonation.

....

....

What could be a detrimental effect of preignition?

....

....

....

Detonation

Symptoms ....

....

....

Causes ....

....

....

....

Remedies ....

....

....

....

Preignition

Symptoms ....

....

....

Causes ....

....

....

....

Remedies ....

....

....

....

Running-on

Symptoms ....

....

....

Causes ....

....

....

....

Remedies ....

....

....

....

COMBUSTION IN A COMPRESSION-IGNITION ENGINE

The combustion process can be described in three distinct stages or phases.

Complete the graph to show the approximate pressure rise in a C.I. engine.

Indicate the three phases of combustion.

Pressure crank angle diagram

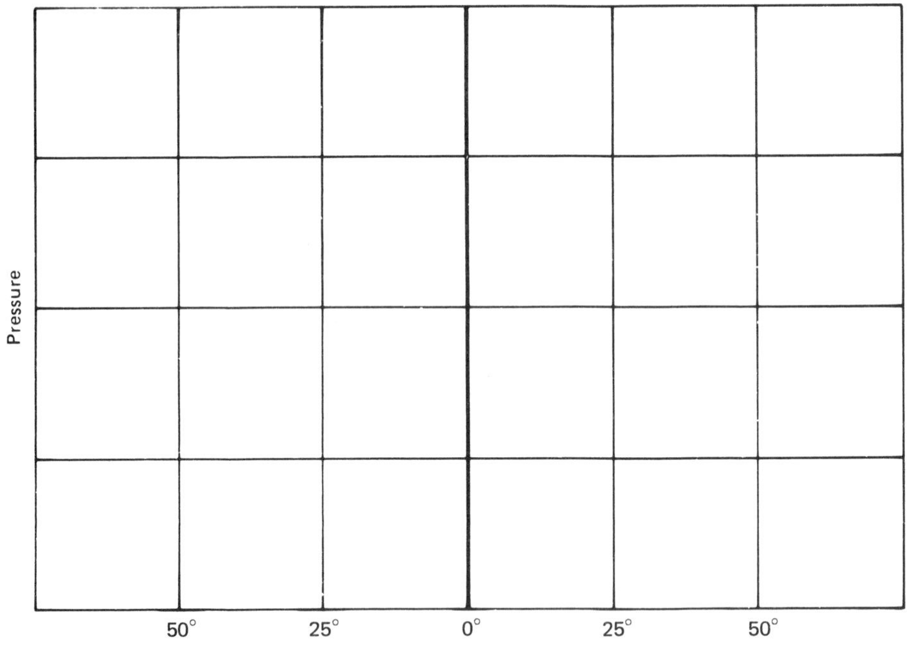

Pressure

50°　25°　0°　25°　50°

....
....
....
....
....

The first phase of combustion is known as the 'delay period'. What is meant by this term?

....
....
....

What affects the length of the delay period?

....
....
....

What is the effect on 'knock' when the delay period is varied?

....
....
....

The amount of fuel supplied and the serviceability of the injector play an important part on the performance output of an engine.

State how the engine performance would be affected by:

(a) excessive fuel

....
....
....

(b) poor atomisation

....
....
....

DIESEL FUEL

This is of the non-volatile type. The approximate chemical composition of a diesel fuel is .... % carbon .... % hydrogen. plus .... % sulphur and .... % oxygen

State the safety advantage of using fuel oil as compared to petrol.

....

....

What is meant by the fuel's 'flash point'?

....

....

How does this differ from the fuel's self-ignition point?

....

....

CETANE NUMBER

The single most important quality of diesel fuel is its ignition quality
This quality is indicated by its cetane number
The cetane rating of a fuel is measured by its ....

....

....

How would diesel knock be influenced by the cetane rating?

....

....

The cetane number of a normal diesel fuel is .... It is determined in a somewhat similar manner to that used to determine the octane rating of petrol. However the reference fuels are hexadecane (i.e. cetane) and alpha-methyl-napthalene.

COUPLES

How do forces produce a couple?

....

....

In a two-cylinder engine a couple is produced that tends to rock the engine in a horizontal plane.

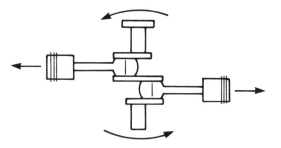

A four-cylinder in-line engine produces at the same time two couples which act in opposite directions.

Indicate the couples acting on the engine shown.

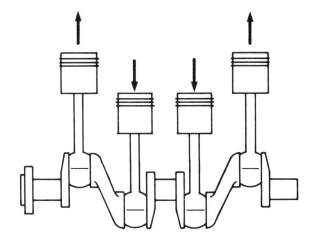

75

ENGINE TESTING

Below is shown one type of apparatus used for measuring the torque and brake power of an engine.

The apparatus is called a ....

The one shown is a .... type

With the aid of these diagrams, describe its basic construction and principle of operation.

Spring
balance

Tachometer

Loading
weight

Sluice gate
control wheel

Sluice plate

Water
inlet

Rotor

Sluice gates

Casing

Sluice
gate
control

....

....

....

....

....

....

....

....

....

....

....

....

....

....

ENGINE POWER

The power measured by the dynamometer is known as the 'brake power'.

Define what is meant by brake power.

....

....

....

'Indicated power' is also a commonly used term.

How is indicated power defined?

....

....

ENGINE TESTING INVESTIGATION

Using an engine suitably mounted to a dynamometer, obtain a series of brake load and fuel consumption values over the engine speed range and from these values calculate the torque and brake power (on this page) and the specific fuel consumption (on p. 82). Use these values to construct graphs which will give an indication of the engine's outputs.

Engine make .... Capacity ....

Describe the engine-testing procedure.

....

....

....

....

....

....

....

....

....

....

....

Calculations

$$\text{Torque} = \frac{\text{Brake}}{\text{Load}} \times \frac{\text{Length of}}{\text{Brake arm}}$$

$$\text{Brake power} = \frac{2\pi \ N \ T}{60 \times 1000}$$

or, using the dynameter constant,

$$\text{Brake power} = \frac{W \ N}{K}$$

Results

Engine speed	
Brake load	
Fuel consumption	

Use fuel consumption values to do the work on p. 82, calculating specific fuel consumption.

L1:104
H1:96

Calculated results

Engine speed	
Torque	
Brake power	

Why is the torque curve at a maximum at a relatively low speed?

....

....

....

What causes the brake-power curve to come to a peak and then rapidly drop?

....

....

....

Problem

A C.I. engine on test, driving a dynamometer whose effective brake-arm length was 350 mm, gave the following results

Speed r/min	1000	2000	3000	4000
Brake Load N	320	360	320	280

Calculate the torque and power at each speed.

78

INDICATED MEAN EFFECTIVE PRESSURE

In order to calculate the indicated power of an engine the average pressure produced during each power stroke must first be determined.

This average pressure is called the ....

Instruments can be used to measure the pressure in the engine cylinder. These pressure valves can then be transferred to a graph shape which is called an 'indicator diagram'. This shows the pressure rise and fall during each stroke and the area enclosed is the work done per cycle.

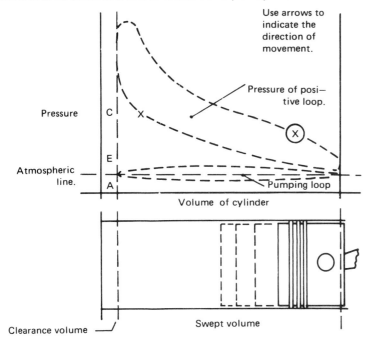

State what is indicated by:

AB ....   DE ....

BC ....   X ....

CD ....   (X) ....

The mean effective pressure is usually found from such an indicator diagram.

The work done (area) may be found, using a planimeter or by the mid-ordinate rule.

The pumping loop is ignored since it is too small to be measured on a normal diagram.

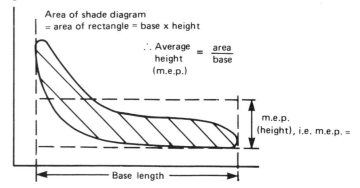

The area of the diagram below has been divided into sections as required when using the mid-ordinate rule.

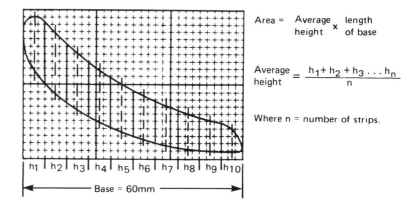

$$Area = \frac{Average}{height} \times \frac{length}{of\ base}$$

$$\frac{Average}{height} = \frac{h_1 + h_2 + h_3 \ldots h_n}{n}$$

Where n = number of strips.

Determine the area of the diagram above in mm^2.

The indicator diagram below shows the shape produced by a typical spark-ignition engine.

Calculate the m.e.p. and the average work done.

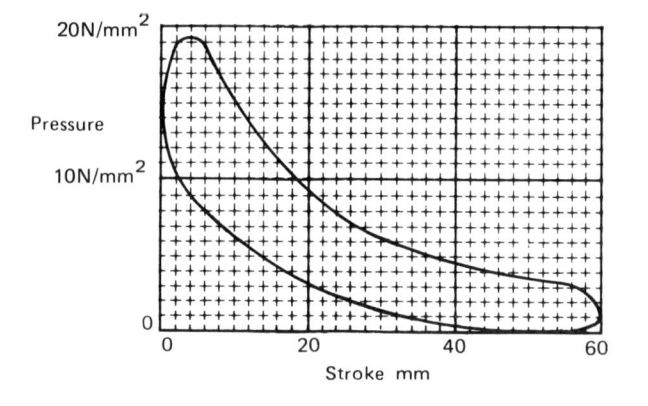

The indicator diagram shows the shape produced by a typical compression-ignition engine.

Calculate the m.e.p. and the average work done.

DIFFERENCE BETWEEN INDICATED AND BRAKE POWER

Indicated power is the actual power developed in the cylinders.
It is the power created before any friction losses occur.
Power must be used to physically turn the engine; this is known as the power that is lost to friction and pumping.

$$\frac{\text{Brake}}{\text{Power}} =$$

1. At 2000 rev/min the indicated power was found to be 32 kW and the power lost to friction to be 3.2 kW. At 5000 rev/min the indicated power was 78 kW and the power lost to friction was 9.4 kW.

   Calculate the brake power at both speeds:

2. The indicated power of an engine was 63 kW and 15% was considered to be power lost to friction. Determine this power loss and the brake power of the engine.

3. The following results were obtained during an engine test.

   Calculate at each speed the power lost to friction and draw a graph to illustrate all three power curves.

Engine speed rev/s	30	40	50	60	70	80	90
Indicated power	15.5	25	35	43	50	53	51.5
Brake power	12.7	21	30	37	43	45.5	42.5
Power lost to friction							

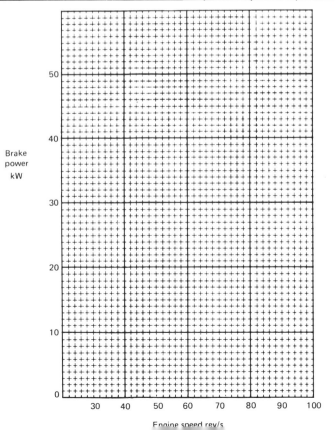

MECHANICAL EFFICIENCY

Mechanical efficiency is the ratio of mechanical power (available to do useful work) compared with indicated power. It is a ratio of the power available in the cylinders (indicated power) and the power at the flywheel (brake power).

Mechanical

efficiency   =

E.g. calculate the mechanical efficiency of an engine whose indicated and brake power were 64 and 51.2 respectively.

Using the values of brake power and indicated power from problem no. 3 on previous page, calculate the mechanical efficiency at each speed.

Complete the table and draw a graph to illustrate mechanical efficiency.

Engine speed	30	40	50	60	70	80	90
Mechanical efficiency							

SPECIFIC FUEL CONSUMPTION

The car owner or fleet operator is usually interested in the fuel consumption of his vehicle in terms of km per litre or miles per gallon. The engine designer, however, is vitally interested in the quantity of fuel consumed per hour for each kilowatt of power developed. This is known as specific fuel consumption.

The fuel consumption is obtained over a range of engine speeds, in units (or converted to units) of kg/h. These values are then divided by the brake power (kW), which is also obtained at these engine speeds. The calculated value is the specific fuel consumption.

The units for specific fuel consumption are ....

or ....

Using the values obtained from the investigation on p. 77, complete the table below and plot the graph of specific fuel consumption on a base of engine speed. Comment upon the results.

Engine speed	
Fuel consumption kg/h	
Brake power kW	
Specific fuel consumption	

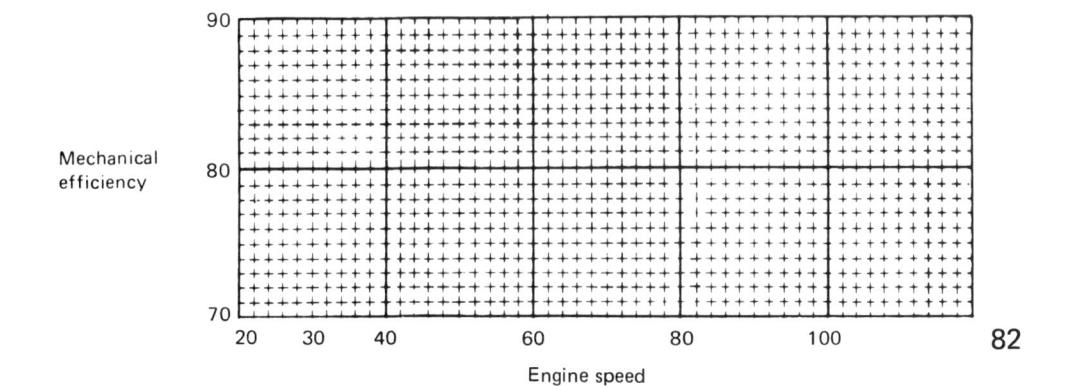

COMPARISON OF PERFORMANCE CHARACTERISTICS

The figures below are typical of readings expected from engine tests of
S.I. and C.I. engines having similar capacities.

Use the appropriate sets of results and show as curves on the graph.

	Engine speed rev/s	10	20	30	40	50	60	70	80
Spark-ignition engine	Torque N m	90	110	104	86	72	59	44	26
	Power kW	14	22	29	35	42	45	46	42
	Specific fuel consumption kg/kWh	0.51	0.41	0.35	0.33	0.35	0.39	0.44	0.50

	Engine speed rev/s	10	20	30	40	50	60
Compression-ignition engine	Torque N m	95	98	97	95	89	83
	Power kW	10	22	33	40	47	46
	Specific fuel consumption kg/kWh	0.37	0.33	0.31	0.31	0.32	0.35

Compare each pair of curves you have drawn and state how they vary with
particular reference as to how the shape of the curves emphasises the
advantages of using either a petrol or C.I. engine.

....

....

....

....

....

....

....

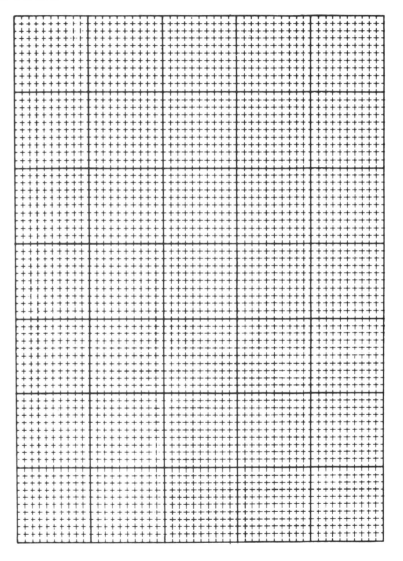

LUBRICATION

APPLIED STUDIES

ENGINE LUBRICATION SYSTEMS

In all modern four-stroke engine lubrication systems the principal bearing surfaces, with the notable exception of the pistons, are supplied with oil under pressure by a positively driven high-pressure pump; other parts are lubricated by splash or mist.

....

....

....

....

....

....

....

....

....

....

....

....

OIL FILTERS

Most engines are fitted with two filters. The first, or primary, filter which is of a coarse wire mesh type and the second, or secondary, filter which is of the replaceable element type.

State the purpose and importance of both these filters.

Primary filter ....

....

Secondary filter ....

....

WET SUMP SYSTEM

With the aid of the diagram below describe how the oil is distributed by pressure splash and mist.

The thick black lines indicate the oil passageways and bearing surfaces.

Name all the main parts.

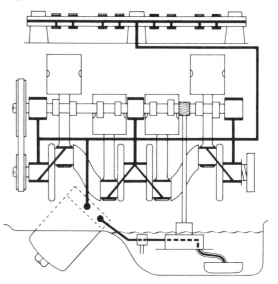

FILTRATION SYSTEMS

There are two types of filtration systems, these are

.... and ....

The system shown above is a .... type.

Describe what is meant by a full-flow system.

....

....

....

The diagrams below are identical: complete the piping details to show a full-flow oil system (diagram 1); a by-pass oil system (diagram 2).

Indicate the direction of oil flow on both diagrams.

1.

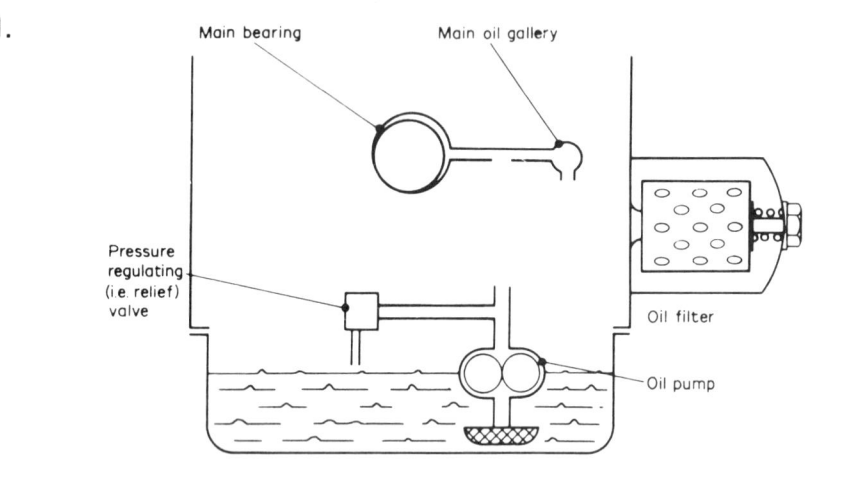

Filter element incorporated in a full-flow system

2.

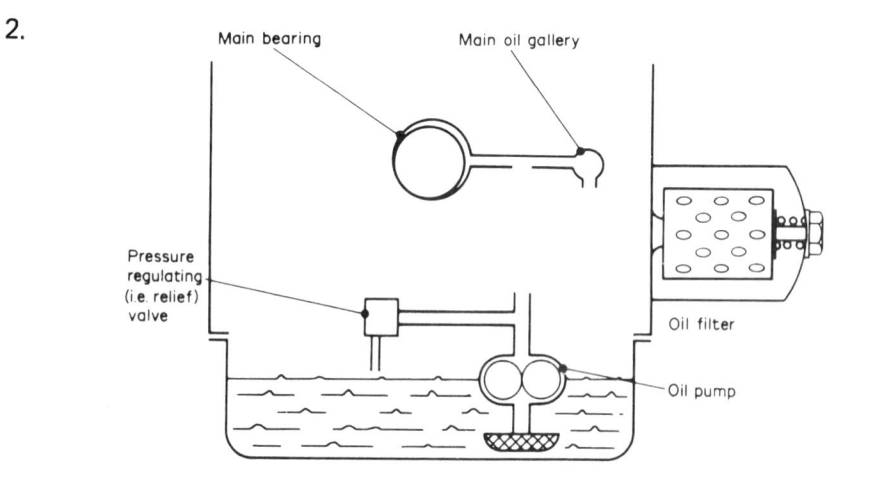

Filter element incorporated in a by-pass system

Describe what is meant by a by-pass system.

....

....

....

State the advantage of both systems.

Full-flow ....

....

By-pass ....

....

Why is the by-pass system rarely used on small engines?

....

....

....

What is the purpose of the full-flow filter by-pass valve?

....

....

....

Describe the basic construction of the filter element.

....

....

....

....

Which way does the oil flow through the element?

....

86

Complete the diagram below to show a sectioned-view of a full-flow oil-filter element; also include the by-pass valve, and show the direction of the flow.

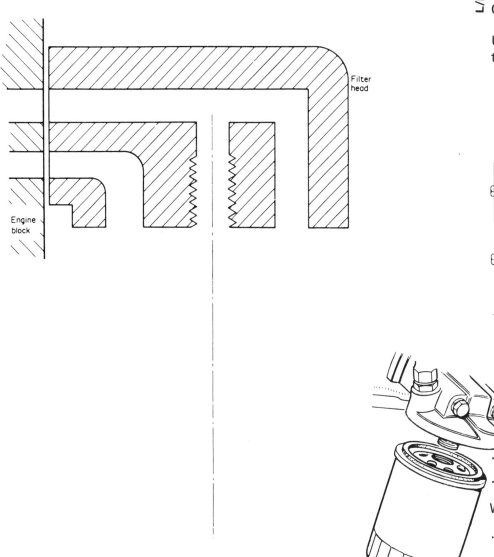

Filter head

Engine block

DRY SUMP SYSTEM

Complete the diagram to show a dry sump system layout.

Using a coloured crayon, show the oil passageways and bearing surfaces for the complete system.

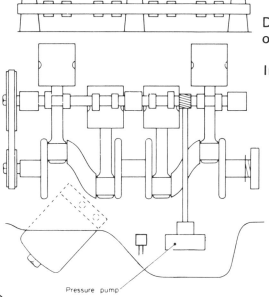

Describe below the system's basic operation.

Indicate direction of oil flow.

Pressure pump

....
....
....

What are the advantages of the dry sump system over the wet sump system?

....
....
....

OIL PUMPS

Examine various oil pumps and complete the diagrams to include the pumping mechanism. Show also the direction of oil-flow and pump rotation.

Eccentric rotor　　　　　　Gear type　　　　　　Eccentric vane

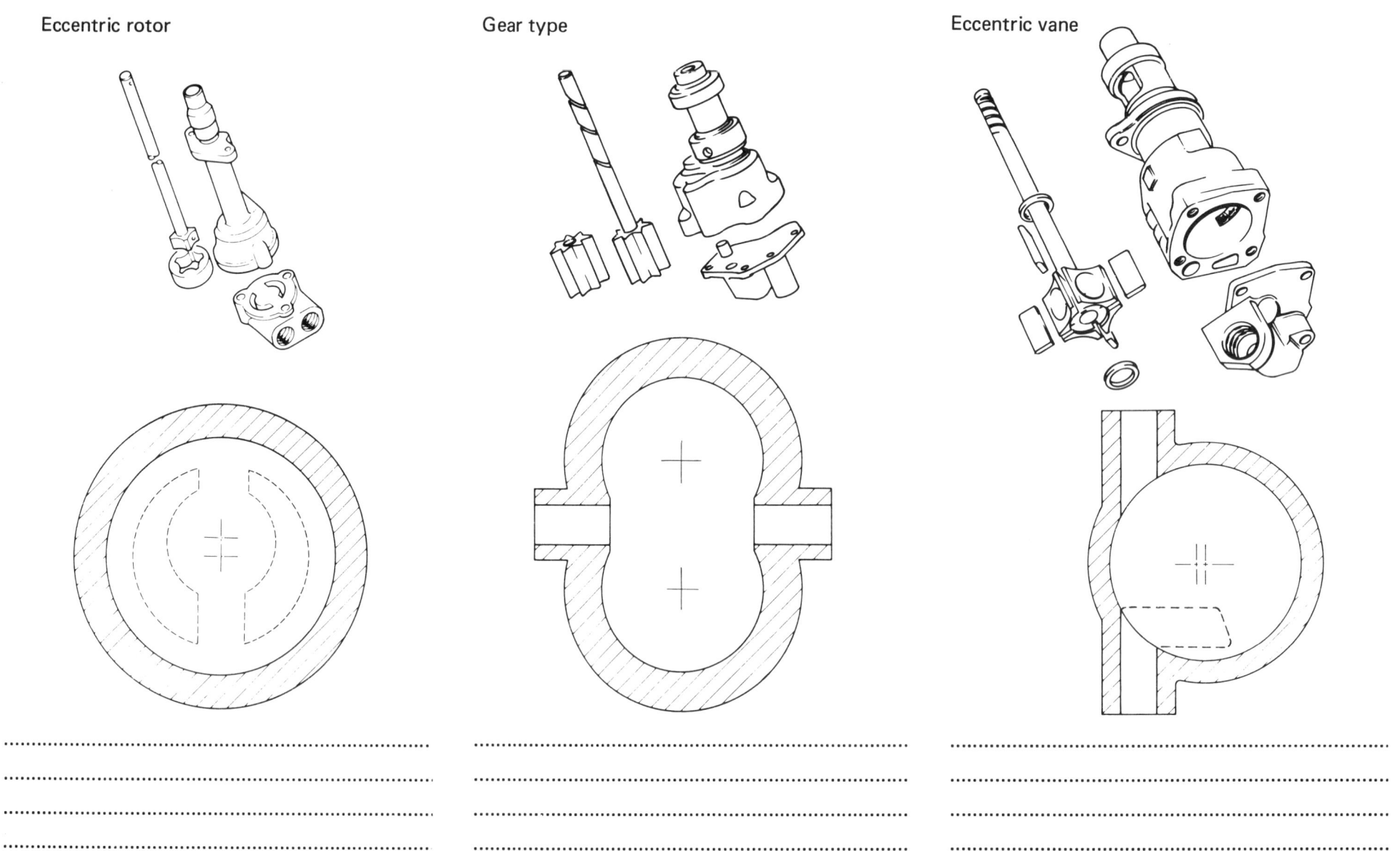

OIL PUMPS

In order to supply oil at high pressure the pump must maintain close working tolerances. This is even more critical when the pump is not submerged in the oil.

Examine an eccentric rotor-type pump and measure the working clearances as shown to determine the pump's serviceability.

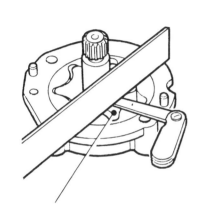

End-float

Lobe clearances

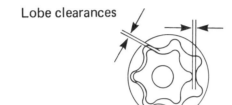

Ring to outer body clearance

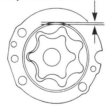

Vehicle make .... engine capacity ....

Clearance (mm)	End-float		Rotor lobe	Ring to outer body
	outer ring	rotor		
Manufacturer's specification				
Measurement taken				

COMMENTS ....
....

OIL PRESSURE RELIEF VALVES

These are usually fitted in the system between the oil pump and the oil filter.

Why is it necessary to fit a relief valve?

....
....
....

Describe their operation

....
....
....
....

Make sectioned sketches of two oil pressure relief valves.

1. Plunger type

2. Ball type

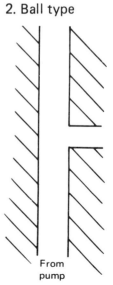

From pump

From pump

89

OIL PRESSURE GAUGES

The gauge below is of the type designed to give a direct oil pressure reading.

Complete the diagram to show the internal components and explain how it operates.

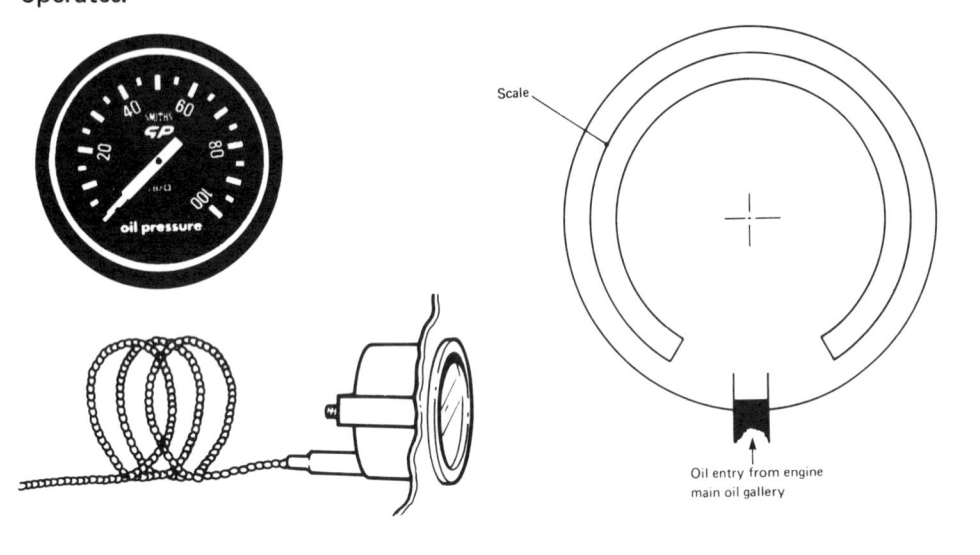

Scale

Oil entry from engine
main oil gallery

....
....
....

An alternative type of gauge would operate electrically in a manner similar to a fuel gauge.

Quote two vehicles that fit a pressure gauge as standard and state typical pressures.

Make	Model	Oil pressure

OIL PRESSURE WARNING LIGHT SWITCH

This is the most common oil pressure indicator arrangement used on mass-produced vehicles.

Complete the sectioned view of the switch and explain how it operates.

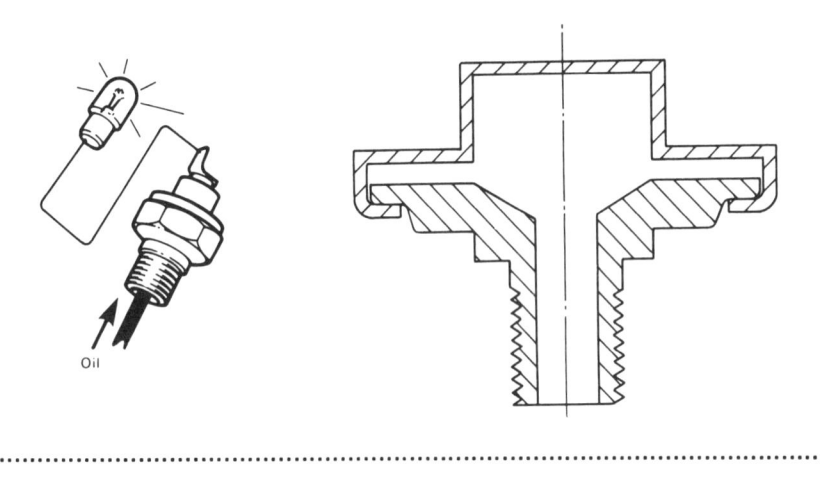

Oil

....
....
....
....

State two positions where the switch may be located on the engine.

....
....

INVESTIGATION

Connect pressure gauge to engine and note pressure when light goes out. Compare with pressure stamped on switch.

Switch operating pressure	Manufacturer's specification

GASKETS AND OIL SEALS

All engine component joints must be suitably sealed. Identify the seals shown and explain why they are constructed from different materials.

....
....
....
....
....

LUBRICATING OIL (TYPE AND GRADE)

When selecting oil for an engine it is important that as well as being of the correct grade it must be the correct type to make it suitable for a particular application.

....
....
....

What is the effect of overfilling the engine with oil?

....
....
....

INVESTIGATION

Inspect various engine specifications in order to complete the following table.

Approximate engine size	S.I. engines		C.I. engines	
	under 1500 cm³	over 1500 cm³	under 5000 cm³	over 5000 cm³
Make/Model				
Cubic capacity				
Oil capacity				
Recommended grade				
Oil change interval				
Filter change interval				

ENGINE OIL SEALING MATERIALS AND METHODS

The material or method suitable for sealing rigid, finely machined surfaces would normally not be suitable for use on pressed steel plate edges or a component that is rotating at speed.

Complete the table below to indicate different materials used as gaskets or seals in the situations stated.

Situation	Material(s)
Sump plug washer	
Oil filter bowl	
Cylinder head	
Sump to block	
Timing cover	
Rocker cover	
Valve stems	
Crankshaft ends	

Sketch a scroll method of oil retention used on a crankshaft and describe its action.

....

....

....

....

....

....

....

PROPERTIES OF OIL

Oil possesses, amongst others, two main properties: body and flow.

Body is the ability of an oil to maintain an oil film between two surfaces.

....

....

....

....

....

....

Flow concerns the property of an oil to spread easily over surfaces and to flow through pipelines and oilways.

....

....

How does the change in temperature affect these two properties?

....

....

Modern oils use many additives to improve the above and other properties. List the requirements that a modern engine oil must meet.

....

....

....

....

....

OIL CONSUMPTION PROBLEMS

1. A vehicle requires topping up with 0.75 l of oil every 1500 km. The oil consumption in km/l would be

2. A driver purchases two 5 l cans of oil and uses 7.5 l for an oil change. He runs the vehicle for 6000 km and uses the remainder of the oil for periodic topping up. Calculate the oil consumption in km/l during this period.

3. An engine burning a lot of oil was topped up daily and over 5 days required:
1.25 l, 0.75 l, 1.5 l, 0.75 l, and 0.25 l.
During this period the total mileage covered was 1260 km.

Calculate the oil consumption in km/l.

4. If an engine oil consumption is 25 000 km/l and the dipstick high level represents 5 l and the low level 3.5 l, how far can the vehicle safely travel before topping up with oil?

5. A vehicle requires topping up with 0.5 l of oil every 950 km. How much oil is used over a 10 000-km period to the nearest 0.5 l.

6. The oil capacity of a C.I. engine is 40 l. When the oil level was checked it required 3.8 l of oil to correct the level. How much oil did the engine contain before topping up?

Express this value as a percentage.

PURPOSE OF A LUBRICANT

All moving components usually have some form of lubricant placed between the surfaces.

The purpose of a lubricant is to: ....

....

TYPES OF LUBRICATION

The types of lubrication may be classed as:

1. .... 2. ....

Lubrication should be based on maintaining a fluid film whenever possible and it is the behaviour of an oil under extreme load and temperature conditions which determines the usefulness of an oil.

When only boundary lubrication is present sometimes the layer of oil is only 1 molecule thick and if the film of lubricant breaks down, due to the high temperatures and pressures that occur in the engine, then overheating, rapid wear and/or seizure could easily result.

The two drawings below each show a much-magnified representation of part of a bearing face. Add a similar sketch above each one, in such a way as to show what is meant by 'full-fluid film' and 'boundary' lubrication.

Full-fluid film lubrication	Boundary lubrication

OILINESS

Oiliness is a term used to describe an oil's body, which means:

....

VISCOSITY

To say that an oil is 'thick' is not sufficiently precise.

....

The viscosity of an oil can be measured by using a viscometer. This measures the time taken for a given quantity of oil to flow through a jet of known diameter at a certain temperature.

The American Society of Automotive Engineers (S.A.E.) have developed a system whereby it is possible to classify oils by their viscosity. The number given, however, signifies the viscosity of an oil only at a certain temperature. The S.A.E. number in no way signifies quality.

Below is shown a table which indicates the flow limits an oil must possess to fall into a certain viscosity range.

S.A.E. viscosity classification for crankcase oils				
Viscosity number	Viscosity range			
	-18°C (poise)		100°C (cSt)	
	min	max	min	max
5W	—	12.5	3.8	—
10W	12.5	25.0	4.1	—
15W	25.0	50.0	5.6	—
20W	50.0	100.0	5.6	—
20			5.6	9.3
30			9.3	12.5
40			12.5	16.3
50			16.3	21.9

Terms 'poise' and 'stoke' are named after Dr J. L. M. Poiseuille and Sir George Stokes.

These are not recognised S.I. units, but are terms still used in the lubrication industry.

....
....
....
....
....
....
....

VISCOSITY INDEX

It is important that oil viscosity remains as stable as possible during changes in temperature. Define viscosity index.

....
....
....
....
....
....
....
....
....
....

VISCOSITY TESTING

Determine the viscosity of an oil and show how the oil's viscosity changes with temperature.

If possible use both a single- and multi-grade engine oil.

Show a sketch of apparatus and describe the method used to determine the viscosity. Plot a graph of the results obtained (Note values may be in Redwood seconds, which is the (pre-S.I.) British unit used to determine viscosity.)

....
....
....
....
....
....
....
....
....
....

RESULTS

Types of oil used   A. ....   B. ....

Temperature °C								
Oil A	seconds							
Oil B	seconds							

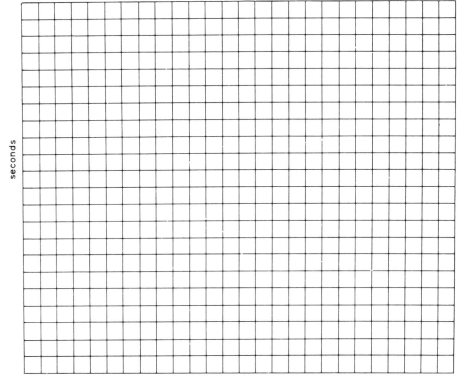

seconds

Temperature in °C

Which oil showed the most stable viscosity? ....

State which grade of oil became least viscous at high temperature. ....

MULTI-GRADE OILS

By improving the viscosity index of an oil (using suitable additives) it is possible for the oil to fall into two viscosity ranges when tested.
State TWO typical multigrade viscosities:

....

Show in graphical form on the special chart below how an oil can be termed multi-grade.

The varying spaces on the graph convert the viscosity curves into straight lines.

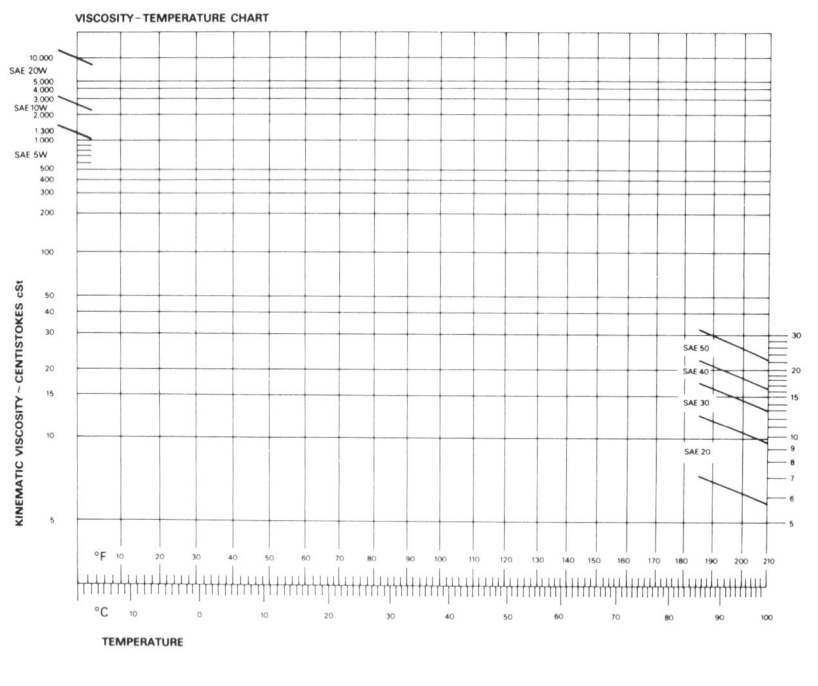

....
....
....
....

What are the advantages of using a multi-grade oil?

....
....
....

Why, if possible, should a 10W30 oil be used in preference to a 20W50 oil?

....
....

OIL CONTAMINATION – THE NEED FOR ADDITIVES

Earlier in this chapter it was mentioned that the engine oil is subjected to extremely high pressures and temperatures. This affects both its viscosity and oiliness. The engine oil also has to contend with other problems, the main one being contamination.

List below the principal contaminants and their source or causes.

Contaminant	Source or cause

Explain the meaning of oil dilution. ....

....

Why is it necessary to change engine oil at regular intervals since other components are filled for life?

....
....

THE USE OF ADDITIVES

Most lubricating oils would be unable to withstand the rigours of severe operating conditions on modern vehicles without the help of additives; these are substances added to an oil in order to improve its performance. An additive may improve the natural properties of a lubricant or confer on the oil properties that it would not otherwise possess.

Listed below are a number of the more common engine lubricating oil additives. By the side of each one write a brief explanatory note.

Viscosity index improvers ....

....

....

Ashless dispersants ....

....

....

....

....

Anti-sludge ....

....

....

Extreme pressure..

....

....

Anti-oxidants ....

....

....

Anti-corrosion ....

....

Rust inhibitors ....

....

'Oiliness' improvers ....

....

Pour point depressants ....

....

....

Anti-foam ....

....

....

State the percentage composition of a typical engine lubricating oil.

Material	Percentage concentration
Base oil	
V.I. improvers	
Ashless dispersants	
Detergents	
E.P. additives	
Anti-oxidants	
Anti-rust	
Pour point depressant	
Anti-foam additive	

FUEL SYSTEMS (PETROL)

APPLIED STUDIES

FUEL SYSTEM (PETROL)

The complete fuel system of a typical modern car consists of:

(a) A fuel tank, usually made from steel, suitably vented and housing some form of fuel-level transmitter. ....
....

(b) A fuel pump, mechanically or .... operated, capable of delivering an adequate supply of fuel to the ....
depending upon engine requirements.

(c) One or more carburettors to atomise ....

(d) Suitable pipelines, mainly made of steel, with some form of flexible connection to accommodate...

(e) Fuel filters usually situated ....
....

TYPICAL FUEL SYSTEM LAYOUT

Label the main components

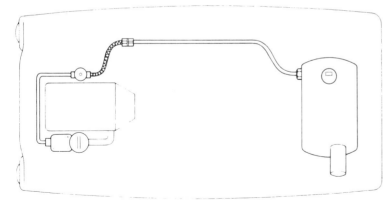

Would this pump be mechanically or electrically operated?

....

INVESTIGATION

Examine the fuel system of a motor vehicle.

Indicate on the sketch below the position of the components of the fuel system and complete the information.

Vehicle make .... Model ....

How is the tank secured in position? ....

Pipe material ....

Make of fuel pump .... Type ....

Make of carburettor(s) .... Type ....

Type of filter(s) ....

How is the fuel tank vented? ....
....

Why is it necessary to provide an air-vent on the tank?...............................
....

FUEL LIFT PUMPS

The fuel lift pump can be mechanically operated from the engine camshaft or electrically operated. In both types of pump the pumping action is provided by a diaphragm flexing within a pumping chamber, into which the petrol is first drawn in and then pumped out, via inlet and outlet valves respectively.

MECHANICAL PUMP

....

....

....

....

....

....

What is the average delivery pressure of this type of petrol pump?.................

How is the delivery pressure governed? ....

....

What happens to the pumping action when the carburettor is full?

....

....

What is the purpose of the smaller of the two springs? ....

....

What is the purpose of the drain hole? ....

....

Dismantle a mechanical fuel pump and complete the drawing below.

Show the inlet and outlet valves and fuel filter diaphragm and return spring, and use arrows to indicate the path of the fuel through the pump.

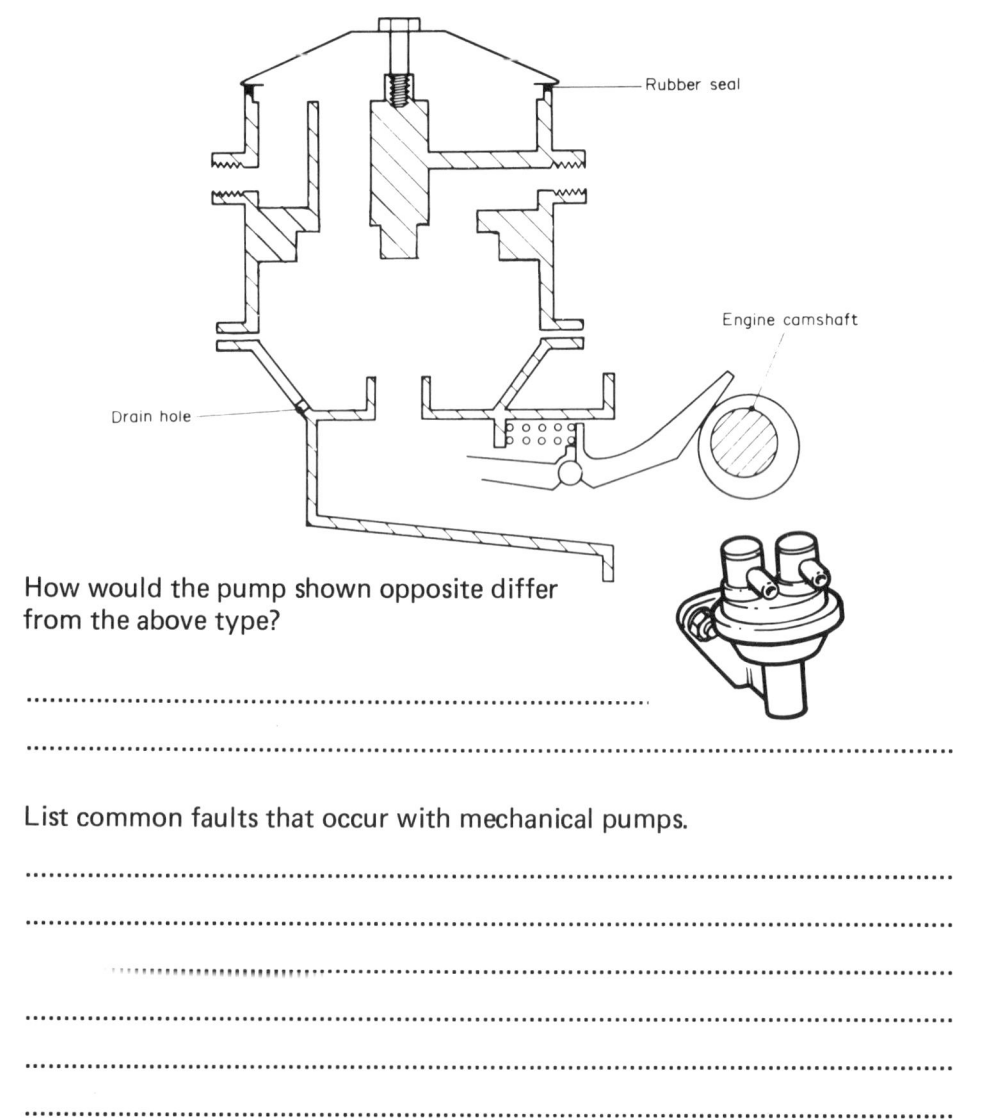

How would the pump shown opposite differ from the above type?

....

....

List common faults that occur with mechanical pumps.

....

....

....

....

ELECTRICALLY OPERATED PUMPS

The diaphragm in the electrical pump is flexed in its chamber by the action of an electromagnet which attracts an iron disc armature attached to the diaphragm. The magnet is controlled by toggle-operated contacts activated by a link rod which is also attached to the diaphragm.

....

....

....

....

....

....

....

....

Describe the action of the pump when the carburettor float chamber is full.

....

....

....

What is the function of the brass rollers?

....

....

One advantage of the electrical pump is that it can be mounted near the petrol tank or indeed in almost any convenient position.

State any other advantages of this type of pump.

....

....

....

Examine an electrically operated pump of a type similar to that shown below and complete the lower drawing by adding: the electrical contacts and toggle mechanism, the inlet and outlet valves and the fuel filter.

Name the major parts on the upper drawing.

Brass rollers

Link rod

Solenoid winding

Armature

List common faults that occur with this type of pump.

....

....

....

THE CONSTANT DEPRESSION (VARIABLE CHOKE) CARBURETTOR

One method of controlling the mixture strength is to vary both the choke area and the effective jet size in accordance with engine speed and load.

A simplified drawing showing the construction of a constant depression carburettor variable choke is shown opposite.

The piston assembly and throttle valve are in the positions occupied when the engine is idling, i.e., the choke area and effective jet size are as small as possible to give the required idling speed.

Why is this known as the constant depression carburettor?

....

....

....

....

....

Describe the action of the carburettor as the engine speed increases.

....

....

....

....

....

....

With this type of carburettor it is extremely important that the mixture strength setting at idling is correct.
Why is this? ....

....

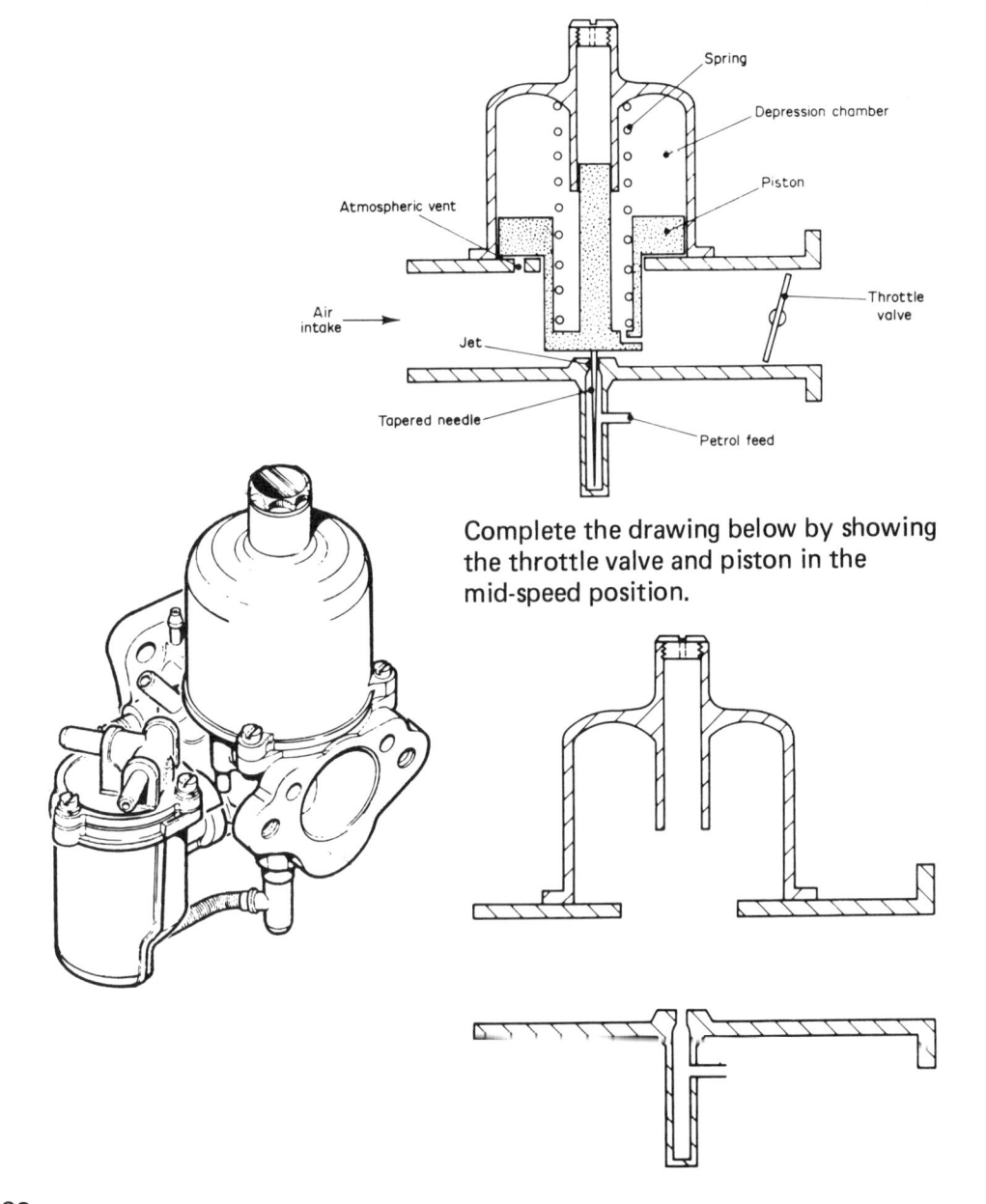

Complete the drawing below by showing the throttle valve and piston in the mid-speed position.

102

S.U. CONSTANT DEPRESSION CARBURETTOR ACCELERATION DEVICE

Examine a constant depression carburettor and complete the drawing opposite to show the type of acceleration device fitted. Describe the operation of the device in the space alongside the drawing.

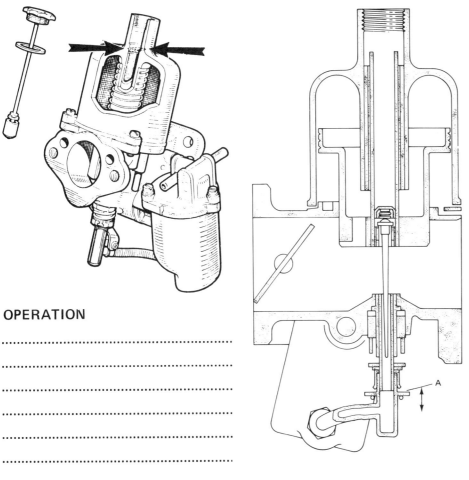

OPERATION

....

....

....

....

....

....

....

Note spring-loaded needle and method of lowering jet.

MIXTURE ENRICHMENT

Show on the diagram right, the position of the choke cable when connected to the mixture enrichment mechanism and name the arrowed parts.

How is the mixture made richer for cold starting?

....

....

....

....

....

....

....

....

....

ADJUSTING MIXTURE

In the carburettors shown on this page the mixture strength and slow running are combined in the same adjustment procedure.

....

....

....

....

....

....

....

....

Raising jet .... mixture.

Lowering jet .... mixture.

ZENITH CONSTANT DEPRESSION CARBURETTOR

The principle of this carburettor is exactly the same as the type described on the previous page. There are, however, some features that are noticeably different.

Describe and indicate on the drawing some of these differing features.

....

....

....

....

....

....

....

....

....

....

....

....

....

....

....

....

....

....

150 C.D.S.

FORD VV (VARIABLE VENTURI) CARBURETTOR

The diagrams show a constant depression carburettor that incorporates a pivoting air valve operating a needle that slides horizontally in the jet.

Name the indicated parts on the external views shown below.

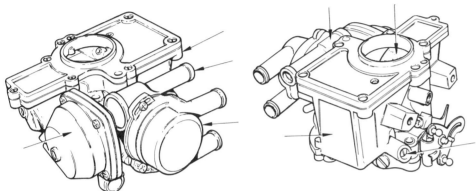

Describe the basic auxiliary system of the carburettor.

....
....
....
....
....
....
....
....
....

The main air-flow system is shown below.

Describe the control of air flow for the engine loadings stated.

Part-load position

....
....

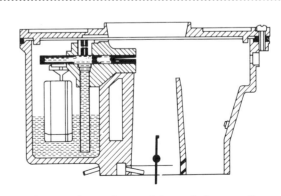

Full-load position

....
....

Complete the drawing to show the position of the needle and air valve at full-load position.

S.U. H.I.F. EMISSION-CONTROL CARBURETTOR

This carburettor operates its main fuel supply system in exactly the same way as described earlier. It is, however, designed to give more precise fuel metering so allowing greater control over emissions at all speeds.

Its main jet design and other features are shown in the drawings.

Describe the action of these features.

JET AND FLOAT CHAMBER LAYOUT

THROTTLE BY-PASS EMULSION SYSTEM

OVERRUN VALVE

A. Fuel supply
B. Air bleed
C. Fuel delivery to jet bridge
D. Commencement of enrichment
E. Maximum enrichment
F. Enrichment outlet
G. Fuel flow through valve

COLD START ENRICHMENT DEVICE

CONSTANT CHOKE CARBURETTORS

The carburettor must control the quantity and proportion of fuel and air entering the cylinders to suit the engine speed and load requirement.

....

....

....

Two systems of compensation are used on constant choke carburettors:

COMPOUND JET SYSTEM

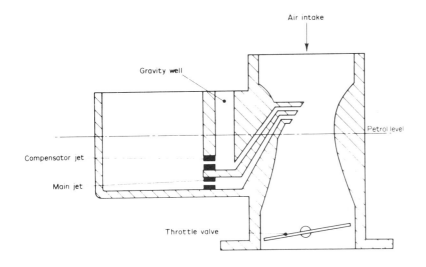

Describe how the combination of the two jets shown above gives good mixture control over a variable speed range.

....

....

....

....

AIR-BLEED SYSTEM

In this system petrol is fed into a well which contains an 'emulsion tube' and an 'air correction jet'. With increase in engine speed the fuel level in the well falls and progressively uncovers air-bleed holes in the emulsion tube. The extra air admitted as a result of this action prevents gradual enrichment of the mixture.

Draw a sectional view of the carburettor venturi to show the action of the air correction jet and emulsion tube.

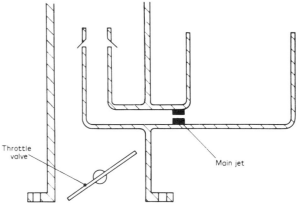

What is the function of the air correction jet?

....

....

What other function does the air-bleed perform?

....

....

How could the mixture strength and degree of correction be altered to suit different engines with this type of carburettor?

....

....

107

IDLING AND PROGRESSION SYSTEM

At low engine speeds the depression around the main jet is inadequate. Therefore the fuel is normally supplied via a pilot jet, through small drillings near to the tip of the throttle valve.

On pre-emission carburettors the layout is as shown below. (For Solex slow-running system, revise p. 63 first-year book.)

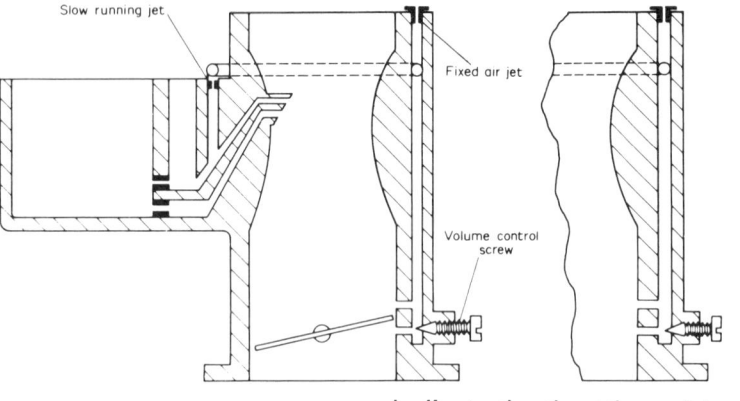

Slow running jet

Fixed air jet

Volume control screw

Use arrows to indicate the path of fuel during idling.

Indicate the throttle position and the path of the fuel during progression.

Why are the fuel outlets for low-speed operation near to the tip of the throttle valve? ....

....

Why is a progression mixture outlet necessary? ....

....

How is the fuel prevented from siphoning through the slow-running system when the engine is stopped?

....

....

SEALED IDLE SYSTEM

Most modern carburettors incorporate an idle system in which the mixture control screw is sealed after setting and then should not require further adjustment. To increase or decrease engine speed an air mixture screw should be turned. This screw is not found on pre-emission carburettors.

Name the parts indicated on the drawing and show the air and fuel flow through the system.

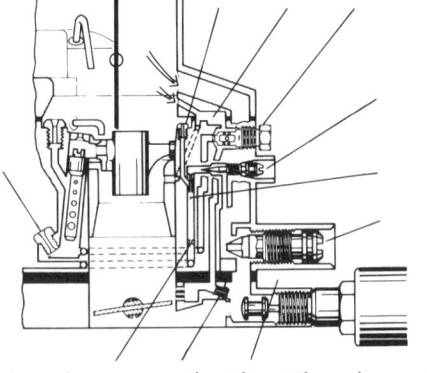

The slow-running mixture by-passes the throttle valve, which must be fully closed during idling. The system can be described in the form of two separate layouts.

1. The basic mixture...

....

....

2. By-pass mixture. ....

....

....

....

COLD-STARTING DEVICES

When the engine is being started from cold the carburettor must supply a richer mixture than is necessary under any other condition. To do this a separate starting device is employed, this device may be either manual or automatic.

STRANGLER STARTING DEVICE
MANUAL

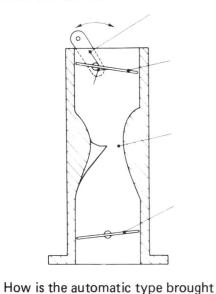

What is the function of the spring?

....

....

Why is the spindle offset?

....

AUTOMATIC

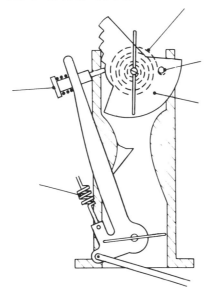

How is the automatic type brought into action?

....

....

How is the strangler progressively opened as the engine warms up?

....

....

....

BY-PASS SYSTEM INCORPORATING A DISC VALVE

This system is completely separate from the main venturi, the air:fuel mixture entering under the throttle butterfly.

The essential component of the system is a disc valve, which is turned by the control lever or is operated automatically.

MANUAL

Show the flow of fuel:air through the system.

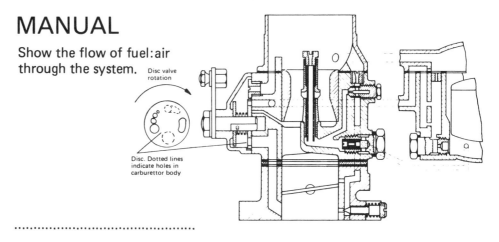

Disc valve rotation

Disc. Dotted lines indicate holes in carburettor body

....

....

AUTOMATIC

The disc may be turned by the action of a bi-metal coil spring which rotates the disc to its off position as its heats up.

State three ways in which an automatic choke may be heated.

....

....

....

Water-heated type.
Name the main parts.

The arrangement shown may be used to operate a disc or strangler valve.

ACCELERATION DEVICES

To give the maximum power required for quick acceleration, the mixture reaching the cylinders should be slightly richer than normal. Without some form of acceleration device, rapidly opening the throttle would produce a weaker mixture than normal, i.e. exactly opposite to what is required.

Why does a rapid throttle opening produce a weak mixture?

....

....

PLUNGER ACCELERATOR PUMP

Complete the drawing of the pump assembly and describe its operation during both rapid and gradual operation.

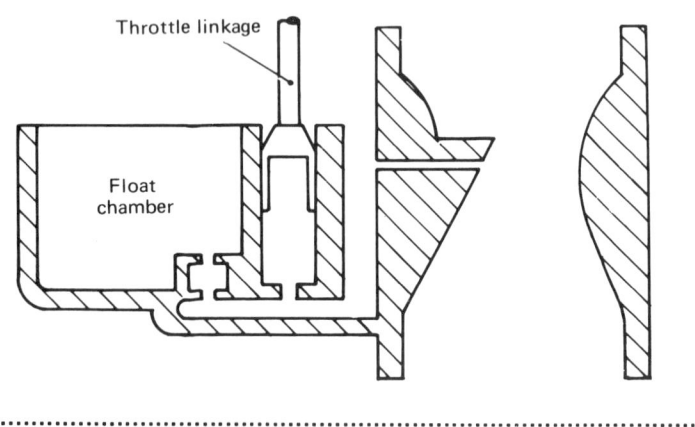

Throttle linkage

Float chamber

....
....
....
....
....
....
....

DIAPHRAGM ACCELERATOR PUMP

The diaphragm provides a pump action in a similar manner to that of a plunger pump.

Show the flow of fuel through the system when the diaphragm is operated.

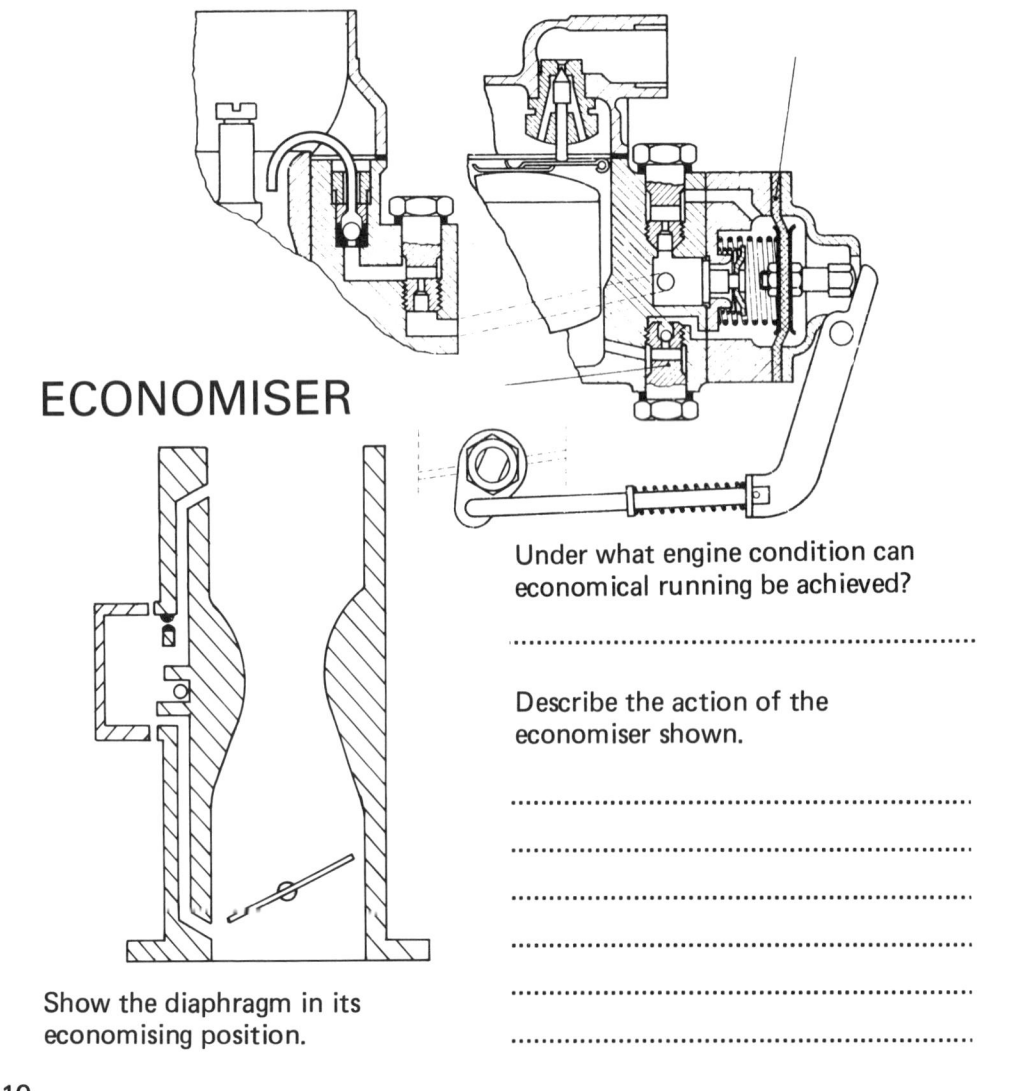

ECONOMISER

Show the diaphragm in its economising position.

Under what engine condition can economical running be achieved?

....

Describe the action of the economiser shown.

....
....
....
....
....

Having studied all the individual systems that create a constant choke carburettor, use this page to revise and to identify the components arrowed.

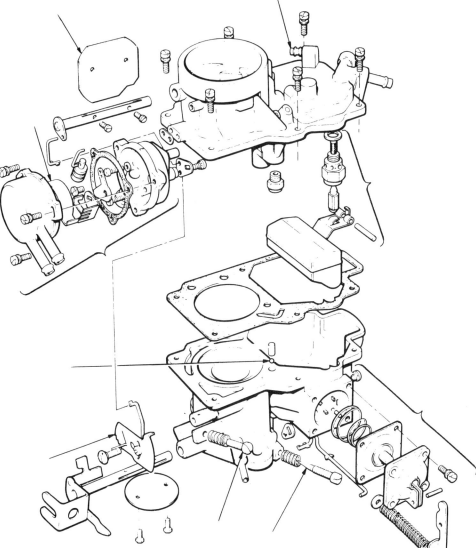

CARBURETTOR FAULTS

List possible carburettor faults from the following symptoms.

Write (C.D.) after the fault if it would occur on constant depression carburettors only.

Excessive fuel consumption ....

....

Idling speed too high ....

....

Poor idling ....

....

Low power output ....

....

Flat spot when accelerating ....

....

Float chamber floods ....

....

Engine loses power at speed ....

....

Difficult starting cold ....

....

Difficult starting hot ....

....

TWIN-CHOKE CARBURETTORS

There are many types of multi-barrel carburettors. The term 'twin-choke' is usually given to those carburettors whose throttles open simultaneously and have a single-float chamber supplying two separate but identical jet assemblies.

Below is shown a twin-choke Weber carburettor

Examine the drawings to trace three of the main operating systems of the carburettor.

Name types of engines that commonly use twin-choke carburettors.

....

....

State the advantages of this type of carburettor when compared with multi-carburettor installations.

....

....

....

....

....

Normal running

....
....
....

Idle speed and progression

....
....
....

Acceleration

....
....
....

PROGRESSIVE CHOKE (TWO-STAGE) CARBURETTORS

Secondary barrel operated directly by a mechanical linkage.

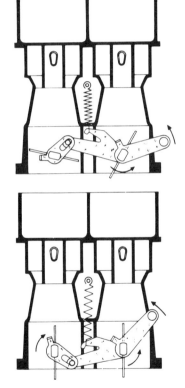

Identify the barrels and, for the throttle positions shown, complete the drawings by showing the petrol spray from the main petrol outlets and add arrows to the drawings to show the air flow through the barrels.

State the essential difference between a progressive and twin-choke carburettor.

....

....

....

In what circumstances are progressive choke carburettors considered necessary?

....

....

....

In the mechanical linkage type (*left*) when the primary linkage reaches three-quarter open position the secondary throttle commences to open and they both reach fully open position at the same time. During normal running, throttle action of the diaphragm type is similar. Describe the throttle action of the vacuum type when the engine is under load.

....

....

....

....

Sketch below a plan view of an inlet manifold for a progressive choke carburettor and show the position of the carburettor barrels in relation to the induction tracts.

Secondary barrel opened by vacuum operating a diaphragm mechanism.

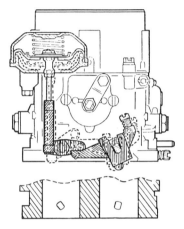

Throttle movement; normal acceleration air speed is sufficient to operate diaphragm.

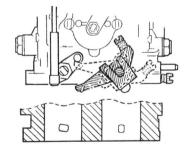

Acceleration under load, air speed to engine is low.

On the lower section of each sketch, show the throttle position related to the position of the linkage on the upper sketches.

MULTIPLE CARBURETTORS

Many high-performance engines use two or more carburettors.
State the reasons for choosing such a layout.

....

....

TUNING MULTIPLE CARBURETTORS

When an engine is fitted with two or more carburettors to obtain satisfactory running, the carburettors must each supply the same mixture strength and the interconnected throttle butterflies must be exactly synchronised.

The actual adjustment of the carburettor requires a certain amount of skill and experience, but nevertheless there is a basic procedure to be adopted when carrying out this adjustment.

On the carburettors shown below, the parts requiring attention during the adjustment procedure are numbered.

Name these parts.

1. .... 4. ....

2. .... 5. ....

3. .... 6. ....

INVESTIGATION

Carry out the adjustment on a twin-carburettor installation on an engine (preferably) fitted with two S.U. or Stromberg (diaphragm type) carburettors.

Use an air flow meter similar to one of those shown below.

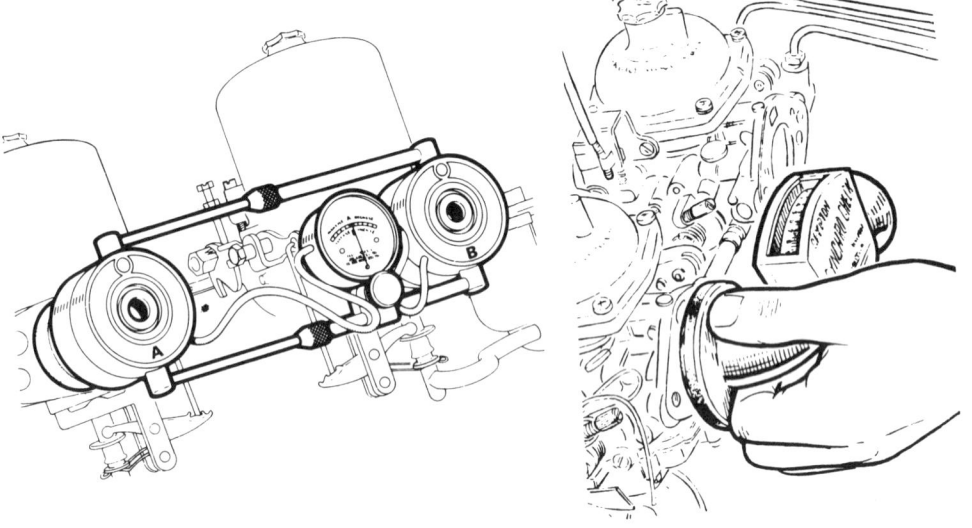

Describe the basic method of synchronising the carburettors.

....

....

....

....

....

....

....

....

AIR CLEANERS AND SILENCERS

The air cleaner and silencer are usually mounted on top of the engine air intake.

The purpose of the assembly is to:

(a) ....

....

(b) ....

....

Name four types of air cleaner:

1...

2...

3...

4...

What would be the effects of a partially blocked air cleaner?

....

....

....

Describe how the air is filtered in types A and B opposite.

A. ....

....

....

....

Examine an oil-bath-type air cleaner and complete the drawing below to show the oil and filter. Indicate also the air flow through the unit.

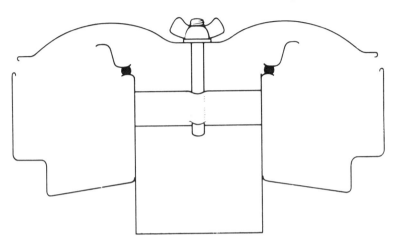

Make a sectional drawing below, of one other type of air cleaner and silencer assembly.

B. ....

....

....

PETROL INJECTION

An alternative method of metering, atomising and distributing the fuel is to use a system of petrol injection.

By using petrol injection, certain difficulties which are inherent in normal carburation systems can be overcome to give advantages of:

....

....

....

....

The basic layout of the Lucas mechanical system is shown below, the principal components being the 'fuel pump', the 'metering distributor unit' and the fuel injectors. Complete the labelling on the drawing and use arrows to indicate the fuel flow through the system.

Vacuum connection

12 volt supply

A simplified drawing of the metering and distributor unit is shown below.

Study the drawing and describe below it, how the fuel is metered and distributed.

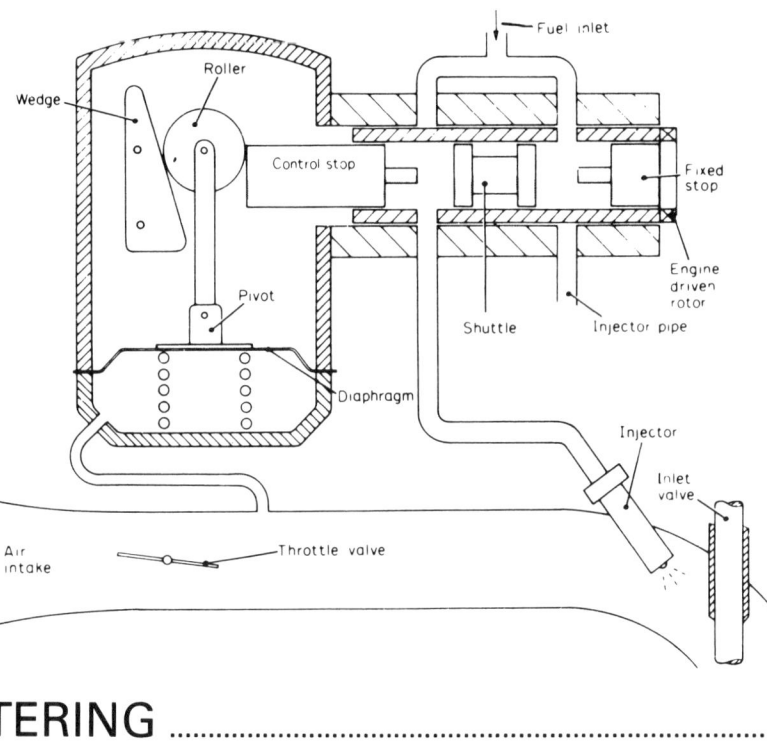

Wedge · Roller · Control stop · Pivot · Diaphragm · Fuel inlet · Fixed stop · Engine driven rotor · Shuttle · Injector pipe · Injector · Inlet valve · Air intake · Throttle valve

METERING ....

....

....

....

DISTRIBUTION ....

....

....

....

116

BOSCH K JETRONIC FUEL INJECTION SYSTEM

In the system shown the fuel is injected continously (K stands for *Kontinuierlich*) instead of intermittently stroke-by-stroke as in the Lucas system.

Examine the layout shown below, name the numbered parts and show the direction of fuel flow when operating normally.

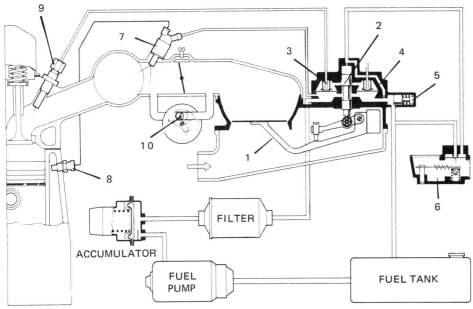

1. ....
2. ....
3. ....
4. ....
5. ....
6. ....
7. ....
8. ....
9. ....
10. ....

Briefly describe the operation of the system.

....
....
....
....
....
....
....
....
....
....
....
....
....
....
....
....
....
....
....
....
....
....
....
....

BOSCH D JETRONIC

The electronic systems of fuel injection are rapidly replacing the mechanical systems. These have a much simpler method of operation.

Name and complete the block diagrams to show:

1. Hydraulic layout

....

....

....

....

2. Electronic layout

The complete basic layout of the electronic system is shown below in schematic form.

Name the main parts.

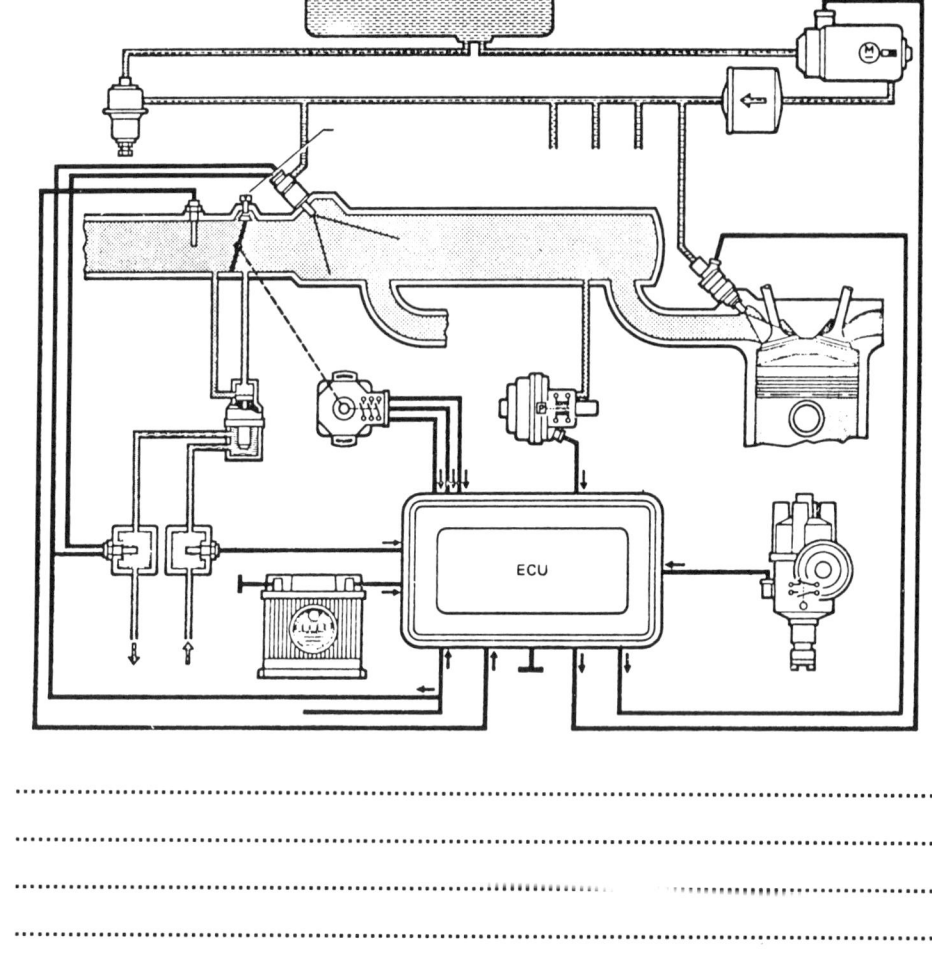

....

....

....

....

....

....

PRESSURE CHARGING

The performance of an internal combustion petrol engine depends to a very large extent upon the density and weight of the charge in the cylinder at the beginning of the compression stroke, i.e. the volumetric efficiency.

State the purpose of pressure charging.

....

....

....

When pressure charging, air is blown into the cylinder to give an increased weight of charge and the engine compression ratio is usually lowered slightly. This increases the 'mean effective pressure' (m.e.p.) throughout the speed range, whilst still maintaining similar maximum compression pressure and maximum engine speed.

Why is it desirable to limit maximum compression pressure and maximum engine speed?

....

What is meant by the term 'mean effective pressure'?

....

MECHANICALLY DRIVEN PRESSURE CHARGERS

There are TWO main types of mechanically driven pressure chargers.

1. ....

2. ....

On the drawing shown at 'A' opposite. Indicate the airflow through the unit in relation to the direction of rotation. Sketch one other type of supercharger at 'B'.

A Type   B Type

ADVANTAGES   # ADVANTAGES

1. ....   1. ....

....

2. ....   2. ....

....

On a petrol engine that uses a carburettor, where is the supercharger normally fitted?

....

Why is it fitted in this position?

....

....

....

How is the supercharger usually driven?

....

TURBOCHARGER

This type of pressure charger is driven by the flow of exhaust gas as it leaves the engine.

Complete the drawing to show the compressor and turbine and describe its basic operation.

....

....

....

....

Diagram shows relative size of rotor blades.

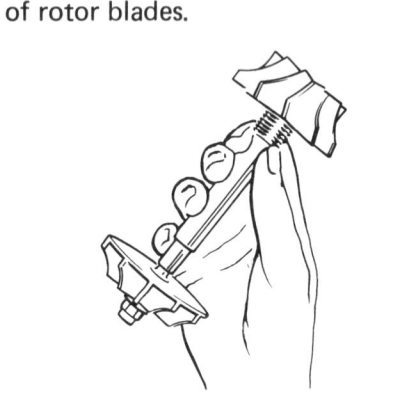

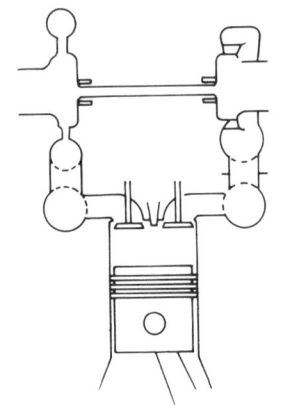

What advantages are gained using this type of pressure charger compared with the other types?

....

....

....

....

Diagrams show possible positions of carburettor and blow-off valve or waste-gate relative to turbocharge.

Blow off valve

Carburettor   Turbine

Compressor

Intake manifold   Exhaust manifold

Carb   Turbine

Compressor   Blow off valve

Intake manifold   Exhaust manifold

....

What is the function of the blow-off valve (or waste-gate)?

....

....

The layout below shows a turbocharger fitted to a petrol injection engine. Indicate with arrows the flow of air through the system.

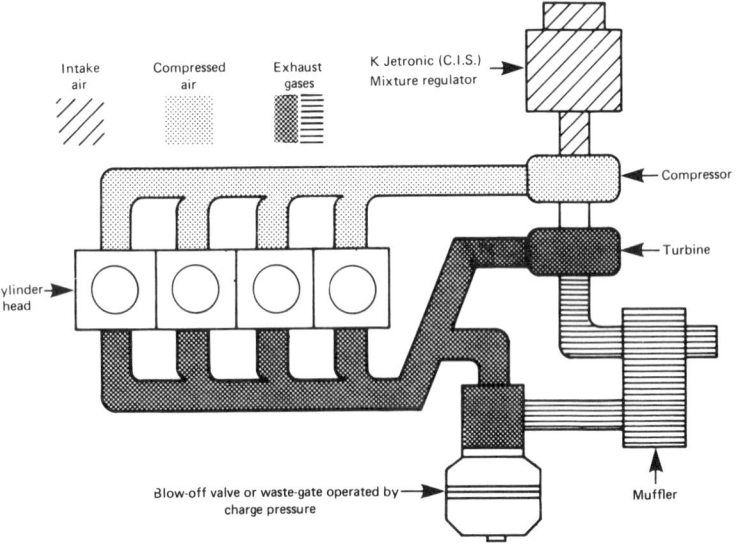

Intake air Compressed air Exhaust gases K Jetronic (C.I.S.) Mixture regulator

Compressor

Turbine

Cylinder head

Muffler

Blow-off valve or waste-gate operated by charge pressure

VACUUM GAUGE

A vacuum gauge can provide an accurate indication of the inlet manifold vacuum of the engine.

What may be assessed from the vacuum readings obtained?

....
....
....
....

Where should the gauge intake be positioned on the engine?

....
....

Poor engine condition would be indicated by too low a vacuum reading.

The diagram indicates some points at which vacuum loss may occur.

List the arrowed items in terms of engine faults.

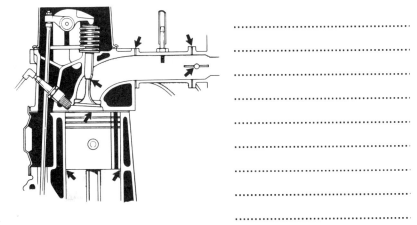

....
....
....
....
....
....
....

FAULT DIAGNOSIS

When diagnosing faults, tests should be carried out under three conditions:

1. .... 2. .... 3. ....

Complete the table to give possible faults and causes.

Test	Correct reading	Possible fault reading	Cause
Engine Cranking	40–50 kPa (12–14 in. Hg) Needle steady		
PCV	Unplug valve reading should		
Idle	50–70 kPa Needle steady Fast idle engines give lower readings		
Acceleration			
Acc. quickly and hold at half-revs.	Falls to 17 kPa then rises.		
Acc. quickly and close throttle.	Rapid fall followed by prompt rise.		
Acc. to near maximum and close throttle.	Fall to zero then momentarily rise above idle.		

EXHAUST GAS ANALYSER

When fault-finding or tuning an engine the use of an exhaust gas analyser will give a good indication as to whether or not correct combustion is being achieved.

There are several types of analysers, the most modern being the infra-red beam type. With this instrument an infra-red beam is passed across the flow of the gas and detects the amount of CO and in some types the amount of HC in the exhaust gas.

CO is ....

....

HC is ....

....

A less expensive meter is the heat-absorption type, which uses a hot wire to burn the passing CO in the gas. The increase in temperature will increase the wires electrical resistance and this will be indicated by an increased CO meter reading on the scale.

What other meter should be used in conjunction with the analyser?

....

Sketch the meter available for use and show the layout of the pick-up tube and condenser when connected to the vehicle.

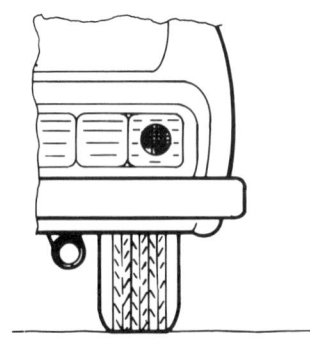

Describe the general engine condition required before the meter is used.

....

....

....

....

....

Describe the meter setting up procedure.

....

....

....

....

State the tests that can be carried out with the meter.

....

....

....

....

Complete the table to show acceptable limits of CO and HC.

Approximate testing speed	Pre-emission-controlled vehicles		Emission-controlled vehicles	
	CO	HC	CO	HC
Idle				
1000 rev/min				
2000 rev/min				

INVESTIGATION

To check the exhaust gas content and performance of an engine connect an exhaust gas analyser, vacuum gauge and tachometer to the appropriate points of the engine.

Warm up the engine to its normal running temperature.

Using the analyser complete the table below.

Vehicle make .... Model ....

Type of carburettor ....

Fault or setting	Analyser reading
Blocked air filter	
Running with air filter removed	
Air leak on inlet manifold	
Rich idling setting	
Weak idling setting	
Incorrect jet fitted	
Choke mechanism partially on	
Analyser reading correct mixture	

Adjust the carburettor and note the effect on the vacuum gauge.

Adjust the mixture screw until the highest *steady* reading is obtained on the gauge; this being an indication of correct mixture strength.

....

Adjust the mixture screw to give a weak mixture and state the effect on the gauge.

....

Adjust the mixture screw to give a rich mixture and state the effect on the gauge.

....

EXHAUST RESTRICTION

What are the effects that a restricted exhaust may have on engine operation?

....

....

....

....

How is this problem turned to an advantage?

....

HOT SPOT

What is the purpose of the manifold hot spot?

....

....

What provision is made for the hot spot when a cross-flow head is used?

....

The sketch shows a thermostatically controlled hot-spot device.

Show the hot control valve in its warm-running position.

Name the parts.

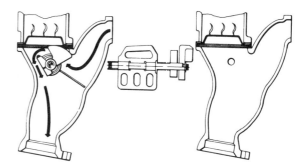

APPLIED STUDIES

PRESSURE

Pressure may be defined as the applied force acting on a unit area

or pressure =

The S.I. unit of pressure is the...

Pressure may also be expressed in two forms.

Gauge pressure and atmospheric pressure.

What is meant by gauge pressure? ....

....

Give an example where such a gauge would indicate:

a positive reading ....

a negative reading...

What is meant by absolute pressure? ....

....

The air pressure on earth is known as..

and is equal to...

This means that air presses with a force of 100 kN on every square metre of the earth's surface.

When an ordinary pressure gauge reads '0' the absolute pressure is ....

Therefore to convert gauge pressure to absolute pressure:

....

An oil pressure gauge gave a reading of 350 kPa.

The absolute pressure would be

The inlet manifold depression when idling was found to be −69 kPa.

An example of pressure change in an engine occurs during the passage of the air:fuel mixture from the atmosphere to its entry to the cylinder.

Shown on the drawing below are typical obstructions met by the mixture as it passes through the system.

Show on the graph typical pressure change created by these obstructions.

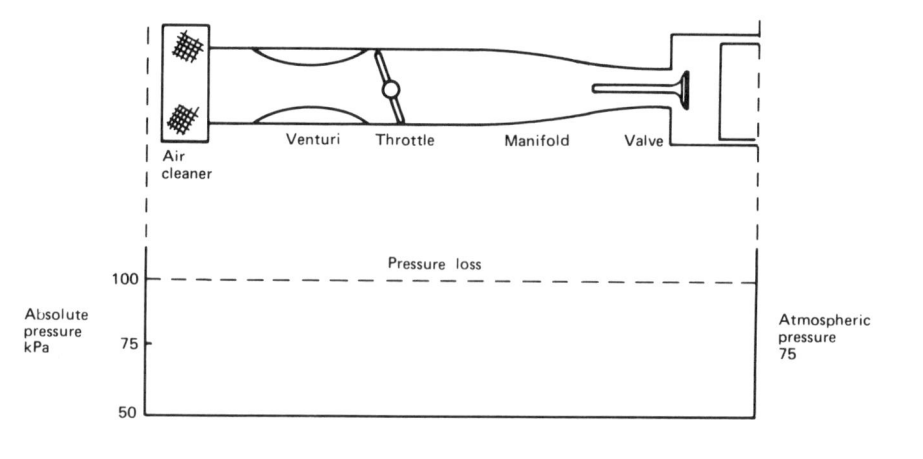

These obstructions prevent the air:fuel mass from completely filling the cylinder. This has the effect of reducing the ....

What other factor affects this efficiency?

....

....

How can this pressure loss be eliminated?

....

What design factors can reduce the pressure loss?

....

....

EXHAUST GAS PRODUCTS

Exhaust gas is the product of the combustion of air and fuel mixture.

Air consists mainly of ....

Fuel consists mainly of ....

State how these products combine to form the exhaust gas.

....

....

....

Below is shown in graphical form the % amounts of CO_2 CO and O_2 in the exhaust gas when an engine is run under maximum load at a constant speed and the mixture strength is varied. Indicate on the graph the position of the chemically correct air:fuel ratio.

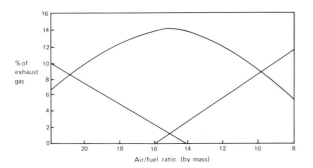

Complete the table to show the percentage of the exhaust gas products when the mixture is as stated.

Mixture	Air:fuel ratio	% CO_2	% O_2	% CO
Correct				
Rich				
Weak				

List the main undesirable exhaust products and state the cause of each.

....

....

....

....

....

....

....

....

....

....

Describe how an exhaust gas analyser may be used to measure the CO content of the exhaust gas in the engine.

....

....

....

....

....

....

....

....

....

....

FUEL SYSTEMS (COMPRESSION-IGNITION)

APPLIED STUDIES

FUEL SYSTEM LAYOUTS

To supply the fuel at high pressure to the injectors, small high-speed C.I. engines use either an in-line jerk pump or a rotary-distributor pump.

The main components of both system layouts are shown on this page.

Investigation

By examining two vehicles each using a different type of injector pump, sketch the piping details. (The coils in the pipes need not be shown.)

Show the direction of fuel flow and name the main components; also indicate the position of the air vents (or vent plugs).

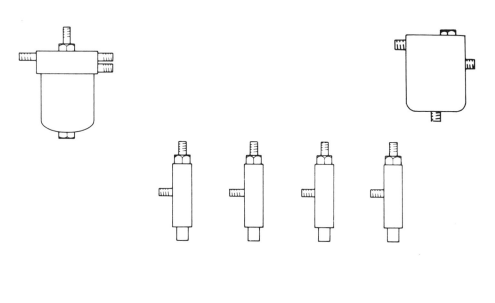

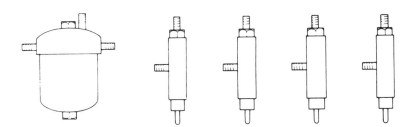

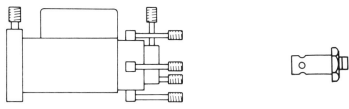

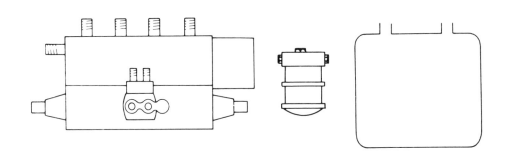

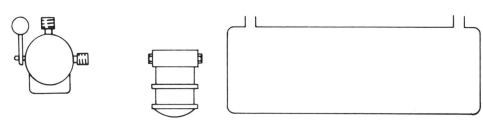

How many air vents or vent plugs are shown in each system?

In-line pump.................................   Rotary pump.................................

FUEL FILTERS

It is extremely important to filter the fuel minutely before it enters the injection pump.

....
....

PRIMARY FILTERS

These usually consist of a bowl which will allow large water droplets and particles of dirt to settle out.

Indicate the direction of fuel flow in the sedimenter shown.

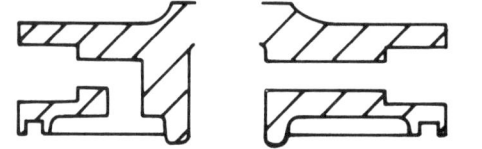

FILTER HEAD
IN
OUT
SEDIMENTER ELEMENT
TRANSPARENT BOWL
SEDIMENTER CHAMBER
DRAIN PLUG

....
....
....

An alternative or addition to the sedimenter would be the fitting of a filter agglomerator.

....
....
....
....
....

Show the direction of fuel flow.

CROSS-FLOW AND DOWN-FLOW FILTERS

The main filter is normally of a paper element type (although fabric elements are still available).

Types of design determine whether the fuel flows down (or up) through the filter (commonly on CAV types) or across the filter similar to a conventional engine oil filter.

....
....
....
....

Complete the sketch below adding labels where necessary. Indicate by means of arrows the direction of the fuel flow.

LIFT PUMPS

Two main types of lift pump are used to supply fuel from the tank to the injection pump on the low-pressure side of the fuel system.

The pump may be driven by the engine camshaft or by the injection pump camshaft.

1. DIAPHRAGM-OPERATED

The diaphragm type is basically similar to those used on petrol engines.

Complete the sketch below to show the position of the valves and show the diaphragm and return spring in diagram (1).

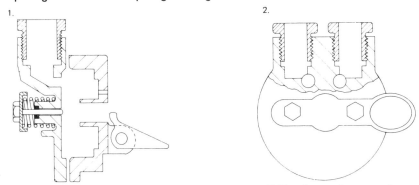

State TWO ways in which the above pump may differ from the one shown on (p. 100)

....

....

....

2. PLUNGER-OPERATED

The plunger-type pump is used to supply fuel at higher pressures. The possibility of air locks is eliminated with this design of pump. The principle is shown opposite.

Inspect a plunger-type pump and complete the schematic drawings below, name the important parts and describe the pump's basic operation.

....

....

....

....

....

....

....

....

....

....

....

....

....

....

....

....

....

....

....

State how air locks are eliminated with this design.

....

....

....

IN-LINE FUEL INJECTION PUMP

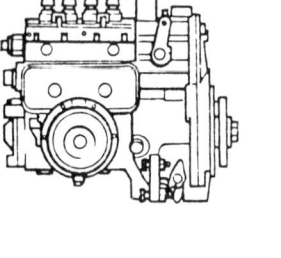

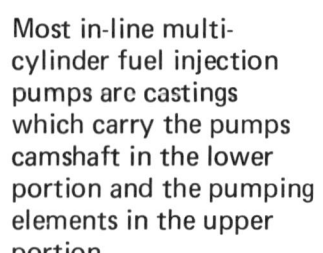

Most in-line multi-cylinder fuel injection pumps arc castings which carry the pumps camshaft in the lower portion and the pumping elements in the upper portion.

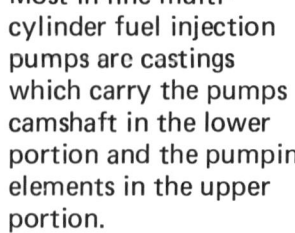

Below are shown two designs of pumping elements; both work on exactly the same principle, but some details are different.

Name the parts indicated.

....

....

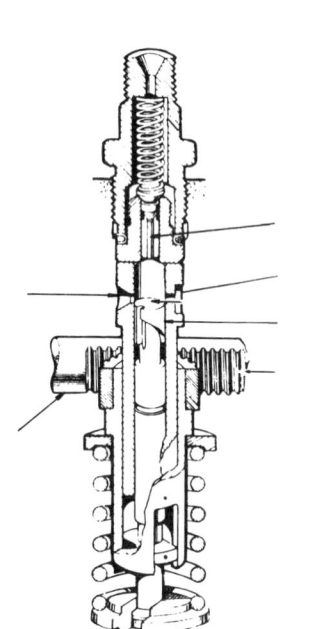

The injection pump must meter exact quantities of fuel, deliver each charge at the correct degree of crankshaft rotation and at high pressure.

Each engine cylinder is fed by a separate pumping element consisting of a constant-stroke, cam-operated plunger.

Add the plunger to each of the sketches below and explain its action.

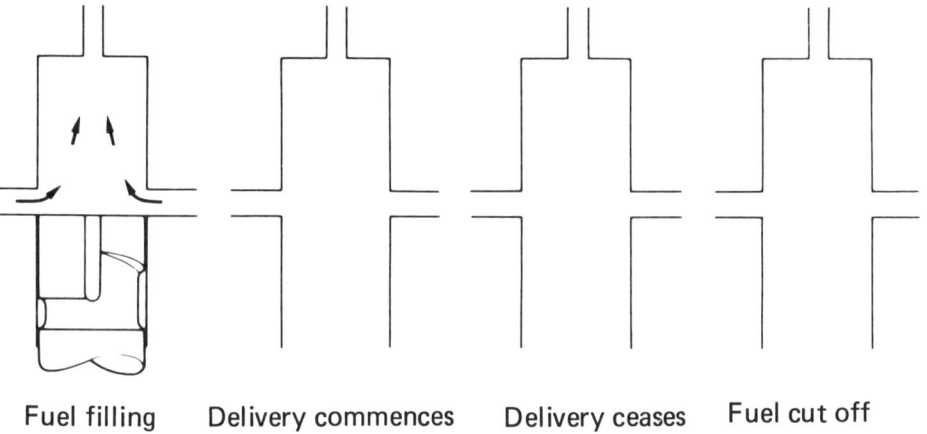

Fuel filling   Delivery commences   Delivery ceases   Fuel cut off

....
....
....
....
....
....
....
....
....
....

DELIVERY VALVE

This valve, or anti-dribble device, performs a dual function. It acts as a non-return valve and as a pressure unloading device.

....
....
....
....

Carefully examine a delivery valve, such as the one shown.
Indicate on the drawing the unloading volume.

Sketch in the space provided below the valve in its open position.

State the effect of a worn delivery valve seat.

....
....

Explain the purpose of the 'unloading collar' machined on the delivery valve.

....
....
....

INVESTIGATION

Remove the side inspection plate of an in-line pump. Rotate the camshaft and note the order in which the plungers lift.

....

When the rack or control rod is moved do the sleeves operate each plunger simultaneously? ....

In which direction does the rack move to increase the fuel supply?

....

How is the delivery of fuel stopped? ....

....

Explain how the movement of the control rod is transferred to the sleeve assembly. ....

....

How are the plunger return spring and cap retained in position?

....

....

Examine a camshaft from such a pump and make a sketch (*on the right*) of the cam shape.

Why is it not the same shape as a poppet valve operating cam?

....

....

What special feature is incorporated into the cam follower?

....

131

PUMP ADJUSTMENTS

It is extremely critical for the efficient performance of the compression-ignition engine that the injection pump maintains precise deliveries of exactly metered quantities of fuel. For this reason facilities for adjustments are provided.

PHASING

This adjustment ensures equi-angular spaced intervals between fuel deliveries. Four element pump .... °. Six element pump .... °.

This test may be carried out on hand-operated equipment or by electronic means. Mount an in-line pump on one of these machines. It is usual to commence with element number 1. Operate the machine noting the interval between deliveries.

Pump make ....

Model..

Show the actual phase angles of the pump on test on the protractor dial.

Using a different colour add the correct phase angles.

Describe the basic procedure.

....
....
....
....
....
....

CALIBRATING

This test ensures that equal quantities of fuel are delivered to each injector for a given control rod setting. The test is carried out by measuring (in graduated tubes) the total delivery, over a number of shots, made by each plunger.

Mount an in-line injection pump on a suitable test machine and conduct a preliminary calibration test noting the quantity of fuel delivered.

Make of pump ....   Type ....

Speed of rotation ....   Number of deliveries ....

Rack setting in mm	Maker's recommended delivery in cm³	Actual delivery in cm³	Corrected delivery in cm³

Describe the basic procedure.

....
....
....
....

EFFECTS OF INCORRECT PHASING AND CALIBRATING

Explain the effect of running a compression-ignition engine with the injection pump out of phase. ....
....

Explain the meaning of the term 'spill cut off'. ....
....

State the effect of a pump delivering unequal quantities of fuel to each of the engine cylinders. ....
....

In which direction has the sleeve to be rotated to reduce the quantity of fuel supplied. ....
....

Observe the lengths of the delivery pipes fitted to the test machine in the previous tests. What is significant about their lengths?
....
....

GOVERNORS

A governor is fitted to the fuel pump; it controls the amount of fuel supplied to the engine and therefore its speed.

In automobile engines the speed is controlled by a combination of the governor's action and the driver's positioning of the accelerator pedal.

State the basic functions of the governor.
....
....

Why would the action of the accelerator pedal alone not control the operation of the C.I. engine?
....
....
....
....
....
....

Describe, in general terms, the control-rod operation of any governor-controlled in-line pump.
....
....
....
....
....
....
....

The above control-rod movements can be controlled by various forms of governors.

1. Mechanical — maximum — minimum speed (older-type vehicles)

2. ....

3. ....

4. ....

ALL-SPEED MECHANICAL GOVERNOR

All mechanical governors rely on the action of centrifugal force attempting to throw a set of spinning weights outwards from their normal position and in doing so, overcoming some form of spring resistance.

MINIMEC–LEAF SPRING GOVERNOR

The type of governor described on this page is commonly found on high-speed diesel engines which may have a cubic capacity of up to 1.5 litres per cylinder.

The governor governs all speeds between idling and maximum speed. Stop screws control the position of the governor at idle and maximum speed.

Label the main parts in the sectional view of the pump below.

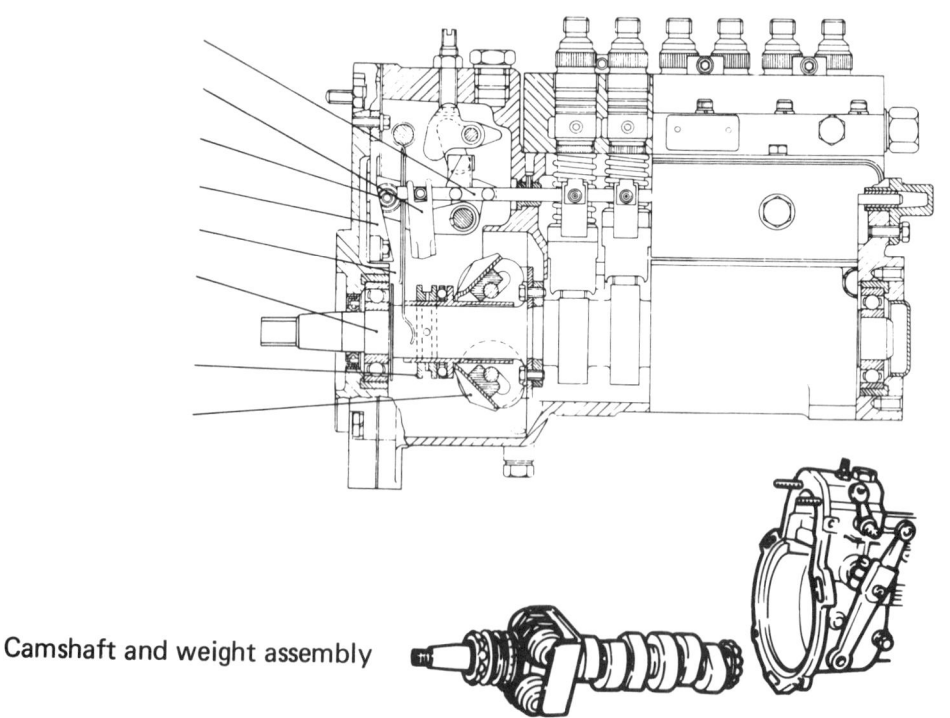

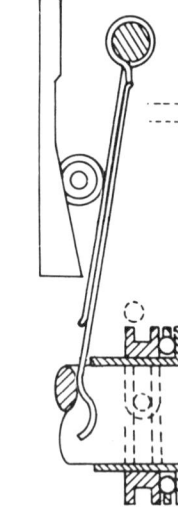

Camshaft and weight assembly

Examine an in-line pump.

Note (1) the operation of the roller weights and (2) the action of the leaf (governor) spring when the throttle lever is operated.

Examine the operation of the governor and name the important parts on the drawings.

....
....
....
....
....
....
....
....
....
....
....
....
....
....
....
....
....
....
....

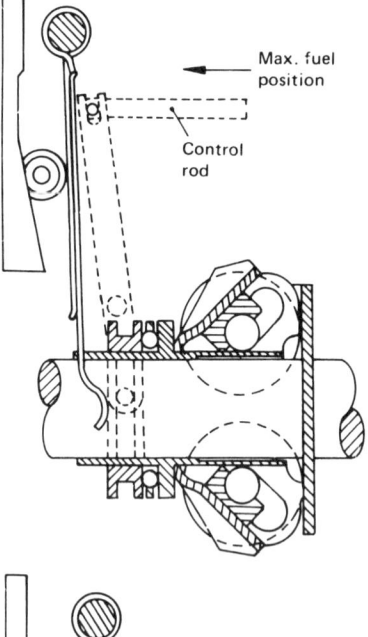

Show position of weights and control rod when pump is operating at idle speed.

MINIMEC – FLYWEIGHT GOVERNORS

This arrangement provides a more sensitive governing particularly at idling speed than the leaf spring governor fitted on identical pumps.

The action of the weights in these governors can readily be seen; indicate this movement by arrows on the first drawing.

State the engine speed related to the flyweight positions shown.

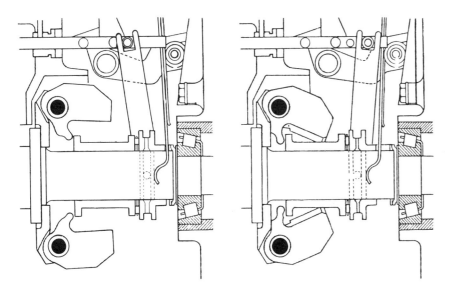

....
....
....
....
....
....
....

ADJUSTMENT OF LINKAGES AND STOPS

The maximum fuel setting and maximum speed adjustment screw are normally predetermined, set and sealed by the manufacturer.

....
....
....
....

The idling speed is adjusted by means of a screw and locknut assembly on the governor unit. Make a sketch in the space below of such an arrangement.

State the maintenance requirements of this type of pump.

....
....
....

List typical faults which could occur with this type of pump.

....
....
....

PNEUMATIC GOVERNOR

This type controls the speed of the engine throughout its range. The governor assembly consists of two main units: a venturi unit mounted on the air intake and a diaphragm unit mounted on the injection pump.

Name the indicated parts.

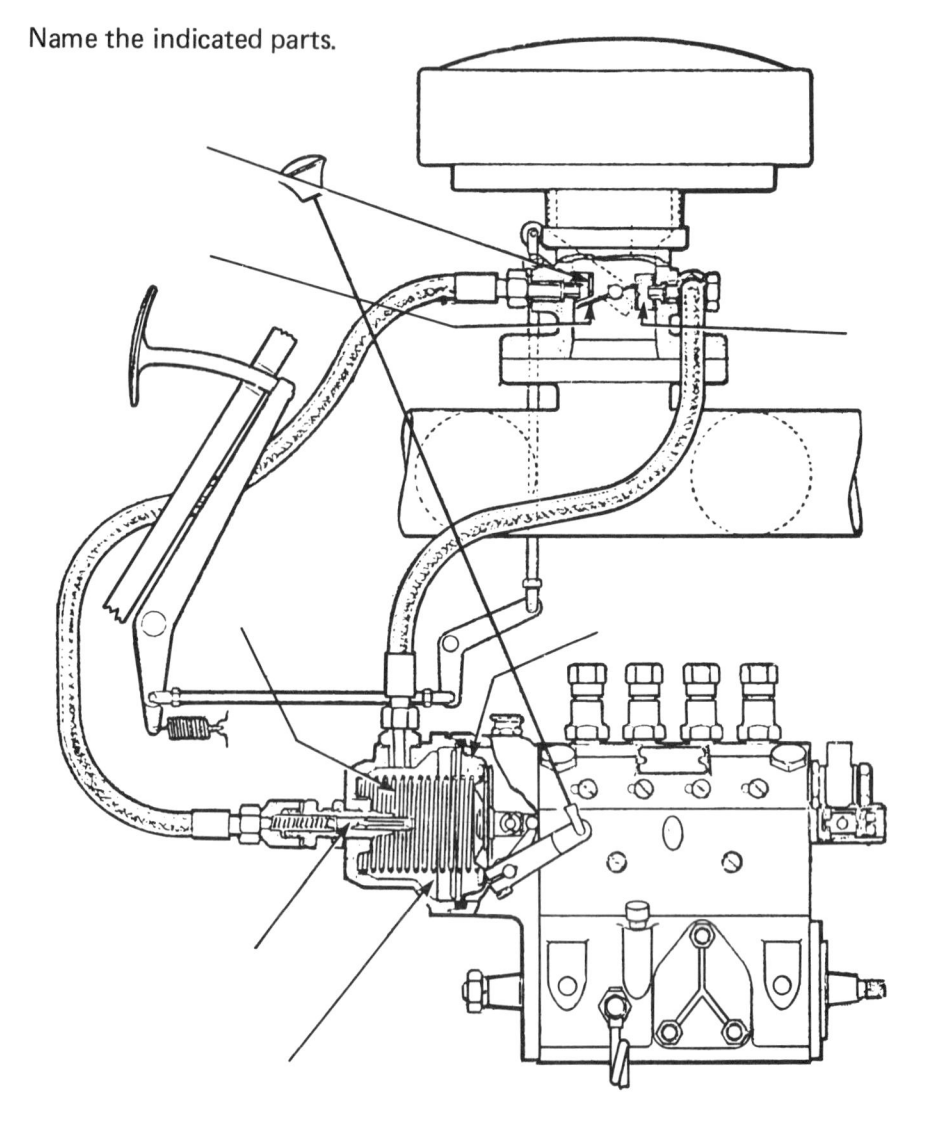

Describe the governor operation.

....
....
....
....
....
....
....
....
....

The idling and maximum speeds on this type of governor can be altered by the venturi unit; the maximum fuel stop is normally pre-set by the manufacturer.

Complete the sketch below to show the control stops on the venturi unit. Label the important parts.

What precaution should be observed when setting up the throttle cable?

....
....
....

PNEUMATIC GOVERNOR INVESTIGATION

Inspect a diaphragm unit and state from which material the diaphragm is made.

....

State the position of the diaphragm and control rod when starting.

....

State the position of the control rod during idling.

....

What is the purpose of the air valve?

....

....

State the maintenance requirements of this type of governor.

....

....

How is it possible to stop fuel delivery with this type of governor?

....

....

Explain the consequences of starting an engine with a vacuum pipe disconnected.

....

....

List typical faults which could make this type of governor inoperative.

....

....

DISTRIBUTOR-TYPE PUMP

The distributor-type pump has one common pressurising and metering device

The pump plungers are operated by a precisely machined cam ring and the rotor unit is driven by the engine.

Complete the diagram below by adding arrows indicating the direction of fuel flow.

Also on the lower diagrams sketch the position of the plungers.

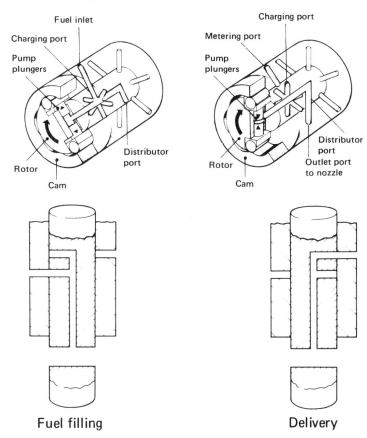

Fuel filling   Delivery

137

The principle of operation of the distributor-type pump is shown on the schematic diagram below; the pump components are shown in the block.

Using four different colours, shade in the pipelines to show inlet, transfer, metering and injection pressures.

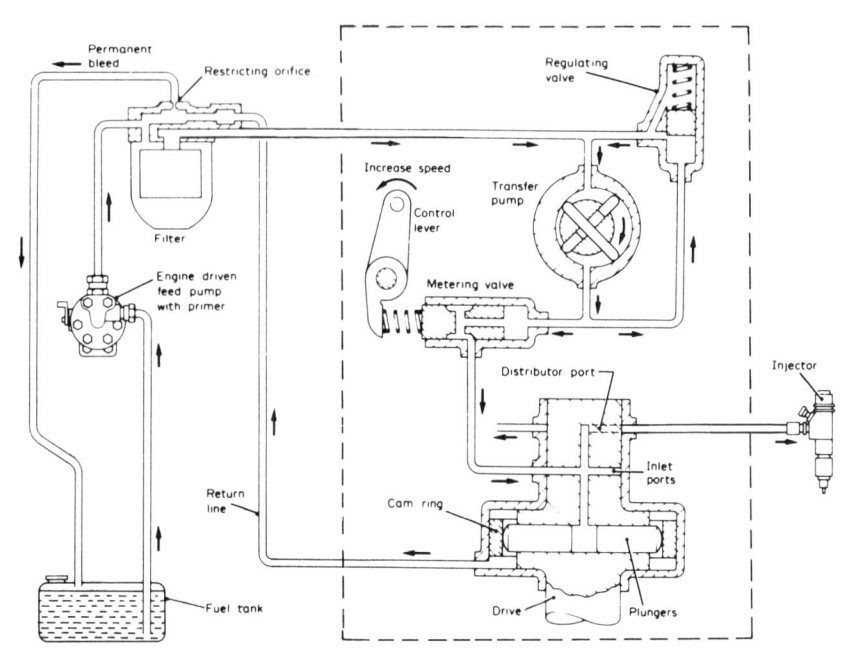

With the aid of the two diagrams describe the operation of the pump.

....

....

....

....

....

....

....

....

With the aid of arrows show the flow of fuel through the pump and name the parts indicated.

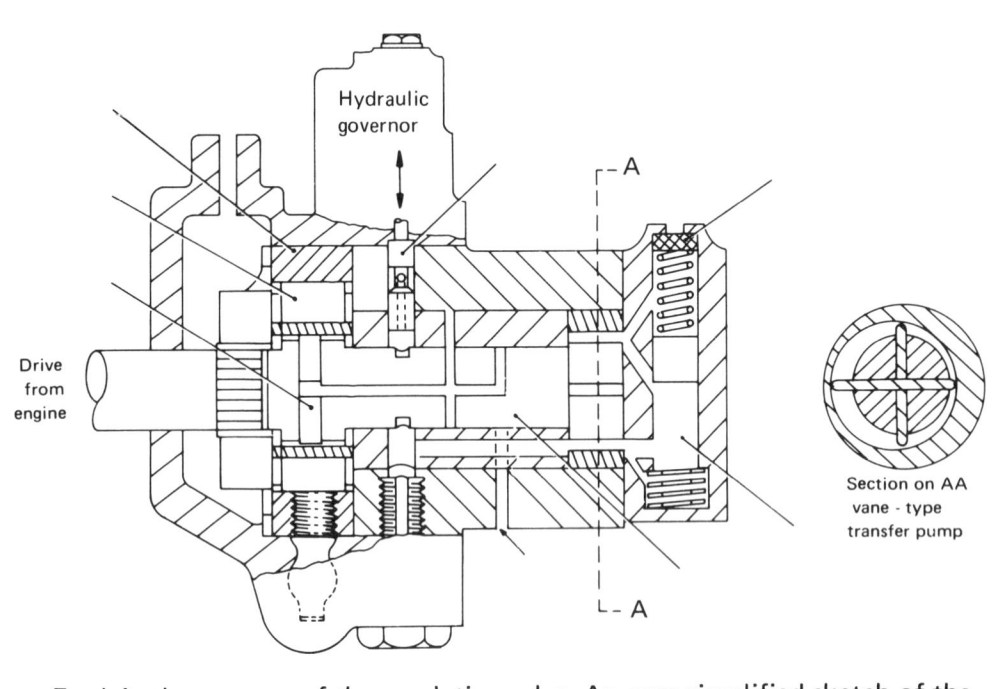

Explain the purpose of the regulating valve. An over-simplified sketch of the valve is shown in both drawings.

....

....

....

....

State how this type of pump ceases to deliver fuel when it is desired to stop the engine. ....

....

....

HYDRAULIC GOVERNOR

The amount of fuel supplied by the metering valve depends upon the balance maintained between the fuel's transfer pressure and the force applied to the governor spring by throttle lever operation.

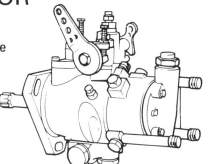

Indicate the governor in the pump drawn above.

Name the parts shown in the drawing below.

(Note the shut-off mechanism.)

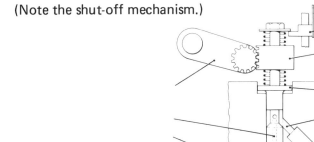

Describe the basic operation of the governor.

....

....

....

....

....

MECHANICAL GOVERNOR

The amount of fuel supplied by the metering valve depends upon the balance maintained between the force created by the rotating weights and the force applied to the governor spring by throttle lever operation.

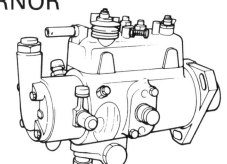

Indicate the governor in the pump drawn above.

Name the parts shown in the drawing below.

(Note the shut-off mechanism.)

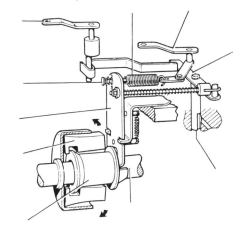

Describe the basic operation of the governor.

....

....

....

....

....

AUTOMATIC ADVANCE MECHANISM

Built into the pump's design is a feature that allows the injection timing to be automatically advanced as the engine speed increases.

Show the direction of rotation of rotor and indicate advance movement on cam ring. Name the important parts.

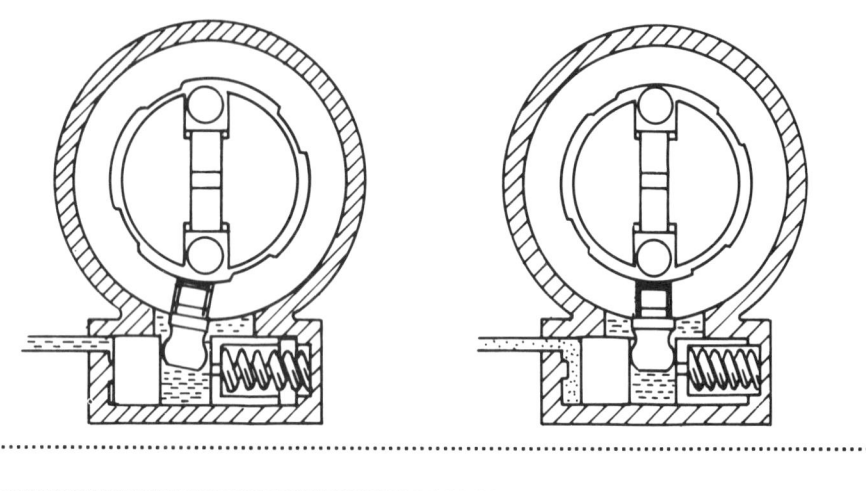

....
....
....
....

A delivery valve is not necessary in this type of pump.

State how pipeline residual pressure is maintained.

....
....

Sketch the shape of the cam ring to show how the pipeline pressure is unloaded to prevent injector dribble.

MAXIMUM FUEL ADJUSTMENT

The amount of fuel supplied during each injection period is dependent upon the stroke of the plungers.

At maximum fuel supply the stroke is limited by the position of the maximum fuel adjustment plate.

The supply is preset when the pump is calibrated before fitting to the engine.

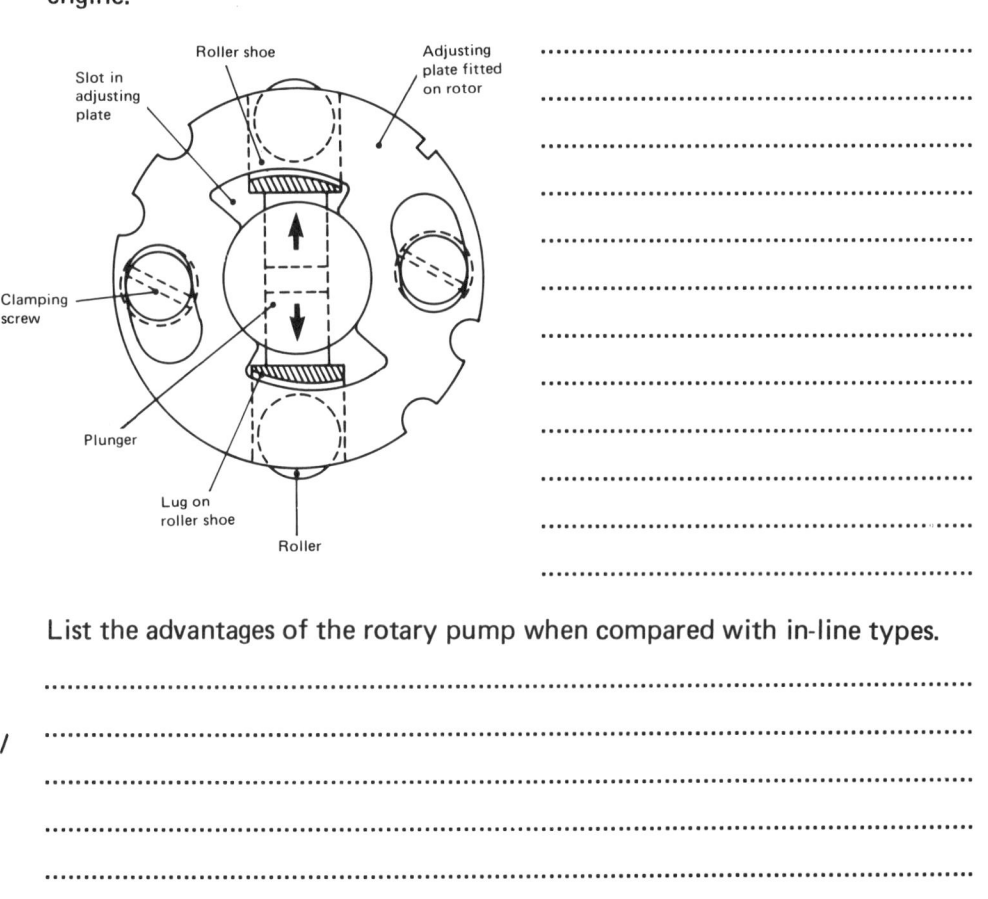

List the advantages of the rotary pump when compared with in-line types.

....
....
....
....
....
....

List common faults which could make the distributor pump inoperative.

....

....

....

....

State what maintenance this type of pump requires. ....

....

....

....

How are the moving parts lubricated? ....

....

....

FITTING AND TIMING INJECTION PUMPS

Both the in-line pump and the distributor-type pump must be fitted to deliver fuel at an exact point of crankshaft rotation. Most distributor-type pumps have a master spline to ensure correct timing. The in-line pump normally has a timing mark which must be in the correct position when coupled to the engine. The flywheel is usually very accurately marked with timing information. However a standard timing procedure can be adopted for most types of pump.

Examine the specifications of several engines and record below the commencement of injection data.

Make and model			
Timing	°b.t.d.c.	°b.t.d.c.	°b.t.d.c.

INVESTIGATION

1. Inspect a typical compression-ignition engine flywheel and sketch in the space below a view of the timing information.

2. Inspect the pump timing marks and location and show a sketch of how they should be set up.

   Engine make .... Type of pump ....

Describe a standard engine/pump timing procedure.

....

....

....

....

....

....

....

....

....

....

INJECTORS

Name the parts of the multi-hole injector shown.

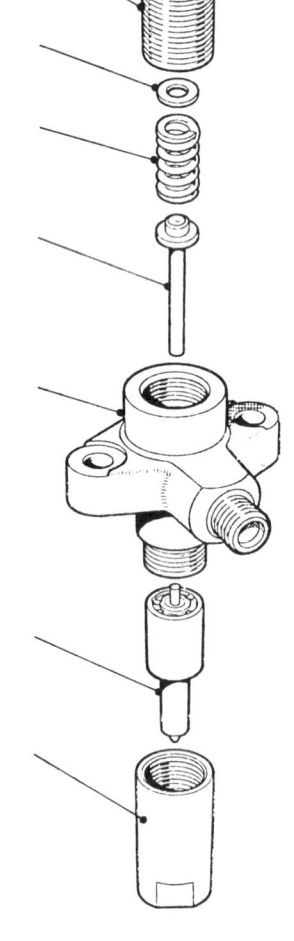

The function of the injector is to deliver fuel in atomised form into the combustion chamber of the engine. The injector normally consists of a body and nozzle assembly.

....
....
....
....
....
....
....
....
....
....
....
....
....
....

Complete the sketch opposite by adding such items as the nozzle body, spring and pin assemblies. Label the essential features.

Explain the purpose of the leak-off facilities provided on an injector.

....
....
....
....

State any special precautions that should be taken when fitting a nozzle nut to the injector body.

....
....
....
....

Explain the meaning of needle lift.

....
....
....
....

Why does the injector buzz when operating?

....
....
....
....

142

TYPES OF NOZZLE

Different designs of combustion chamber demand special nozzles. The two main designs in common use are the hole-type and pintle-type.

The type of engine normally using a:

(i) hole-type injector is ....

(ii) pintle-type injector is ....

HOLE-TYPE

These have an internal needle exposing from one to several holes.

....

....

....

....

Complete the sketches below by adding the nozzle or nozzle pin where necessary.

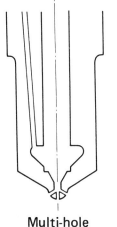

Multi-hole

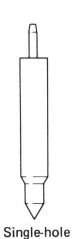

Single-hole

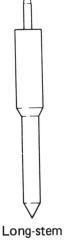

Long-stem

PINTLE-TYPE

This type gives a hollow cone-shaped spray. It is possible slightly to delay the delivery of the main spray by a modification to the pintle, thus giving a more gradual pressure rise and smoother running.

....

....

....

Complete the sketches below by adding the nozzle or the nozzle pin where necessary.

Pintle

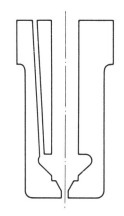

Delay

In the space below complete the sketch by adding the pintaux nozzle pin.

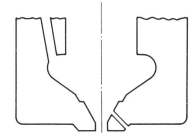

Explain the purpose of the pintaux-type of nozzle.

....

....

....

....

....

....

INJECTOR FAULTS

List the main symptoms that would indicate faulty injectors on a compression-ignition engine.

....

....

....

....

Complete the fault diagnosis table below. These faults may be observed when testing injectors.

Fault	Possible causes
Failure of nozzle to buzz whilst injecting	
Excessive leak-off	
Too high nozzle-opening pressure	
Too low nozzle pressure	
Dripping nozzle	
Nozzle blued	
Distorted spray form	

INVESTIGATION

State the safety precautions which must be exercised using a pressure tester.

....

....

....

Conduct pressure tests on the injectors of at least four different engines or on four differing types of injector.

Make of injector	Type of nozzle	Actual injection pressure	Recommended injection pressure

Using a pressure tester observe the spray form of a hole-type and pintle nozzle. Sketch the outlines below.

144

INVESTIGATIONS

Describe one method of adjusting the pressure setting of an injector.

....
....
....
....
....
....
....
....

Describe one method of quickly locating a faulty injector whilst *in situ* on the engine.

....
....
....
....
....
....
....
....

Examine several hole-type and pintle nozzle injectors and state the hole diameters observed. What maintenance do the holes require?

....
....
....

Type of nozzle				
Hole diameter (mm)				

TESTING

This should include pressure tests, leak tests and spray form.

Pressure test different types of injectors and record your observations below.

Make of test equipment ....

Model ....

Type of test oil used ....

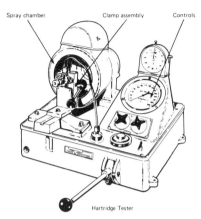

Spray chamber   Clamp assembly   Controls

Hartridge Tester

No.	Injector type	Back leak-age time(s)	Pressure	Seat tightness	Spray form	Recommended pressure
1						
2						
3						
4						

State your observations and recommendations in respect of each injector tested.

1. ....
....

2. ....
....

3. ....
....

4. ....
....

PRESSURE – TIME INJECTION SYSTEM

This system was developed by Cummins Engine Company Inc. Its operating principle is based on the fact that changing the *pressure* of liquid flowing through a pipe changes the amount coming out of the open end (that is, injector); the length of *time* during which the injector is 'open' is controlled mechanically. Hence the name pressure-time system.

The basic mechanical linkage is shown below in diagrammatic form.

The drawing below shows the basic layout of the Cummins P.T. external fuel line system. Name the three major fuel lines and use arrows to indicate the direction of fuel flow.

Describe how the fuel is supplied to the ....

injector. ....

How are the high pressures needed to inject the fuel into the combustion space created?

....

....

....

....

....

....

....

....

....

....

....

With this system about 80% of the fuel delivered to the injectors is returned to the tank. What are the reasons for this?

....

....

....

....

....

....

....

....

Injector

Fuel flow

P.T. fuel pump

Accessory drive

Camshaft gear

Crankshaft gear

Inlet connection

Drain connection

Shut down valve

Fuel filter

P.T. fuel pump

....

....

....

....

The drawings below show the nose of a P.T. injector at various stages of operation. Explain opposite each one what is occurring.

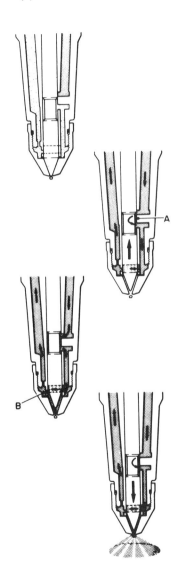

....
....
....
....
....
....
....
....
....
....
....
....
....
....
....
....
....
....
....
....
....
....
....
....
....

What are the relative merits of the P.T. system compared with the more conventional fuel injection systems?

ADVANTAGES

....
....
....
....
....

DISADVANTAGES

....
....
....
....

Sketch below a greatly enlarged view of the nozzle tip, showing clearly the hole 'B'.

COMBINED PUMP – INJECTOR

With this system the injector takes on the dual function of metering and injection. It is commonly known as the General Motors system.

Name all the important parts of the combined pump—injector unit shown. Indicate the direction of rack movement to increase and decrease fuel.

Sketch an alternative injector nozzle.

Open type

How does the open type injector differ from the closed type?

....

....

....

....

....

....

Describe the basic operation of the system.

....

....

....

....

....

With the aid of the diagrams explain the injector operation.

Lower port

Upper port

Top of stroke

Start of injection stroke

End of injection stroke

Bottom of stroke

....

....

....

....

....

....

COLD STARTING

Cold starting compression-ignition engines can be a problem due to initial heat losses. This is particularly so with the indirect injection type. The difficulty can be overcome in several ways.

HEATER PLUGS

These plugs have at the cylinder end a small coil of wire which glows at red heat when switched on.

....

....

....

Examine a heater plug circuit and sketch the wiring diagram in the space below.

Was the circuit wired in parallel or series? ....

State the voltage power rating of each heater plug. ....

State the amperage of the fuse in the circuit. ....

DECOMPRESSION DEVICES

This device is fitted to ease the load on the starter motor and allow the engine to rotate faster. The increased speed results in higher cylinder temperatures. A device is fitted to hold the valves clear of the seats.

....

....

....

....

Complete the sketch showing the decompressor cam in relation to the rocker arm.

State the approximate lift of the cam, sufficient to overcome compression.

....

List the disadvantages of this device.

....

....

....

ETHER SPRAY

A further version of cold-starting aids is the use of an ether spray.

....

....

....

INDUCTION MANIFOLD HEATERS

These are fitted in the inlet mainfold.

One type consists simply of a coil of resistance wire.

....

....

The most common form of manifold heating is to use a hot wire-coil which directly operates a fuel supply.

....

....

....

....

In the space below complete the sketch of a manifold heater incorporating a fuel vaporiser.

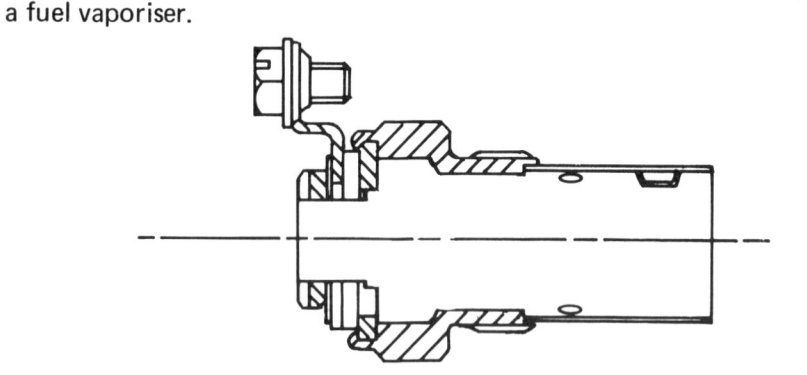

Explain how the fuel is supplied and stored for use in the heater unit. (See second fuel system layout, p. 127.)

....

....

....

EXCESS FUEL DEVICE

Most in-line pumps are fitted with a device which when actuated allows the control rod to move to a position in which excess fuel for easy starting will be delivered.

The push-button type is the simplest arrangement.

Show the plunger in the sketch below in its non-operational position.

....

....

....

....

....

....

....

....

....

....

Alternative types which are 'cheat proof', i.e. the button cannot be 'tied down', may be used when required.

What are the legal requirements of cold-starting devices?

....

....

....

....

PRESSURE CHARGING C.I. ENGINES

Pressure charging increases the air pressure supplied to the inlet manifold of the engine. This has the effect of:

....

....

Why is the pressure charging of C.I. engines much more popular than the pressure charging of petrol engines?

....

....

....

....

....

TURBO SUPERCHARGER

This makes use of the energy in the engine's exhaust gases which are directed on to a turbine which is made to rotate the supercharger.

The drawing shows the basic layout of a turbocharger.

Label the parts and show the direction of rotation and of gas flow.

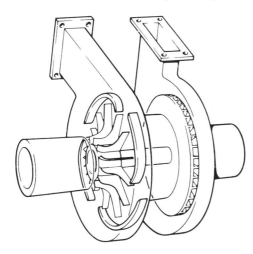

In a basic turbocharged system the compressed air is forced directly into the engine (see p. 120).

On certain C.I. engines an intercooler or charge cooler is used to cool the air after passing through the turbocharger.

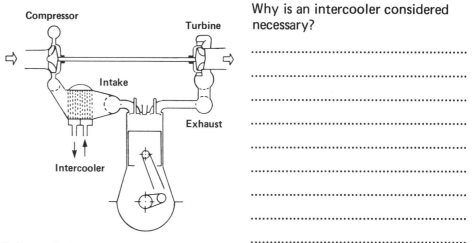

Schematic layout

Indicate, using arrows, the air flow on both diagrams.

Perkins T6.3543 air-to-air intercooling

Why is an intercooler considered necessary?

....

....

....

....

....

....

....

....

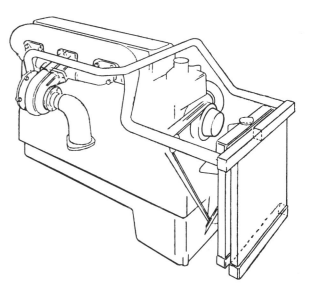

Physical layout of engine and charge cooler

151

The rotational speed of the turbocharger is governed by the exhaust gas flow. What possible speeds may be attained by turbochargers?

....

In order to maintain these high speeds efficient lubrication is critical. Describe how the turbocharger is lubricated and what periodic maintenance is required?

....

....

....

....

What precautions must be observed when starting and stopping a turbocharged engine?

....

....

....

What are the relative merits of this arrangement compared with mechanically driven superchargers?

ADVANTAGES ....

....

DISADVANTAGES ....

....

....

EFFECTS OF AIR IN THE FUEL SYSTEM

It is essential that any air in the fuel is removed before the fuel enters the high-pressure part of the fuel pump.

What effects may occur if any air is in the system?

....

....

....

....

....

....

Explain how it is possible for air to find its way into the fuel system.

....

....

....

....

....

Examine a distributor-type pump and state where provision for bleeding is made. ....

....

Inspect an in-line pump and note where provision is made for bleeding.

....

List the precautions to be taken before bleeding an injection pump.

....

152

IGNITION

APPLIED STUDIES

CONVENTIONAL COIL IGNITION SYSTEM

Complete the semi-pictorial wiring layout of a four-cylinder coil ignition circuit — shown below.

State a suitable firing order, ensure that the secondary circuit wiring complies with this firing order and draw a rotor in the distributor cap firing on number ONE cylinder.

Name the various parts.

Firing order ....

The coil ignition system is basically divided into two circuits, PRIMARY and SECONDARY (low and high tension).

State the basic function of each main component.

....
....
....
....
....
....
....
....
....
....

IGNITION COIL

Examine a sectioned or dismantled ignition coil and describe its construction (preferably after completing the sketch on the next page).

....
....
....
....
....
....
....
....
....

Explain the operation of the ignition coil.

....
....
....
....
....
....
....

Complete the sketch to show the wiring and core arrangement.

RESISTOR COILS

Most modern cars use a special low-voltage coil which during normal running is connected in series to a ballast resistor.

This ballast resistor may be placed in various positions in the ignition's primary circuit.

....
....
....
....
....
....
....
....
....

What is the reason for using a ballast resistor in the primary circuit?

....
....

Draw a primary circuit diagram to show the positioning of a ballast resistor.

To starter

Explain how such a low-voltage coil and ballast resistor operate.

....
....
....
....
....
....
....

What effect would leaving the ignition switched on (without the engine running) have on the ballast resistor?

....

DISTRIBUTOR UNITS

The purpose of a distributor can be divided into three distinct roles. These are to:

....

....

....

....

CONTACT BREAKER POINTS

In relation to contact breaker points at what instant in their operation is the spark at the plug produced.

....

Describe the operation of the contact breakers.

....

....

....

....

....

Some manufacturers fit 'ventilated'-type points or sliding points. How do these points differ from the normal type and what are their advantages?

....

....

....

The contact breaker points are positioned on the base plate of the distributor. The movable point is operated by the cam.

Examine two types of distributors used on four-cylinder engines similar to those shown in the upper drawings and complete the plan views to show the contacts in their assembled position.

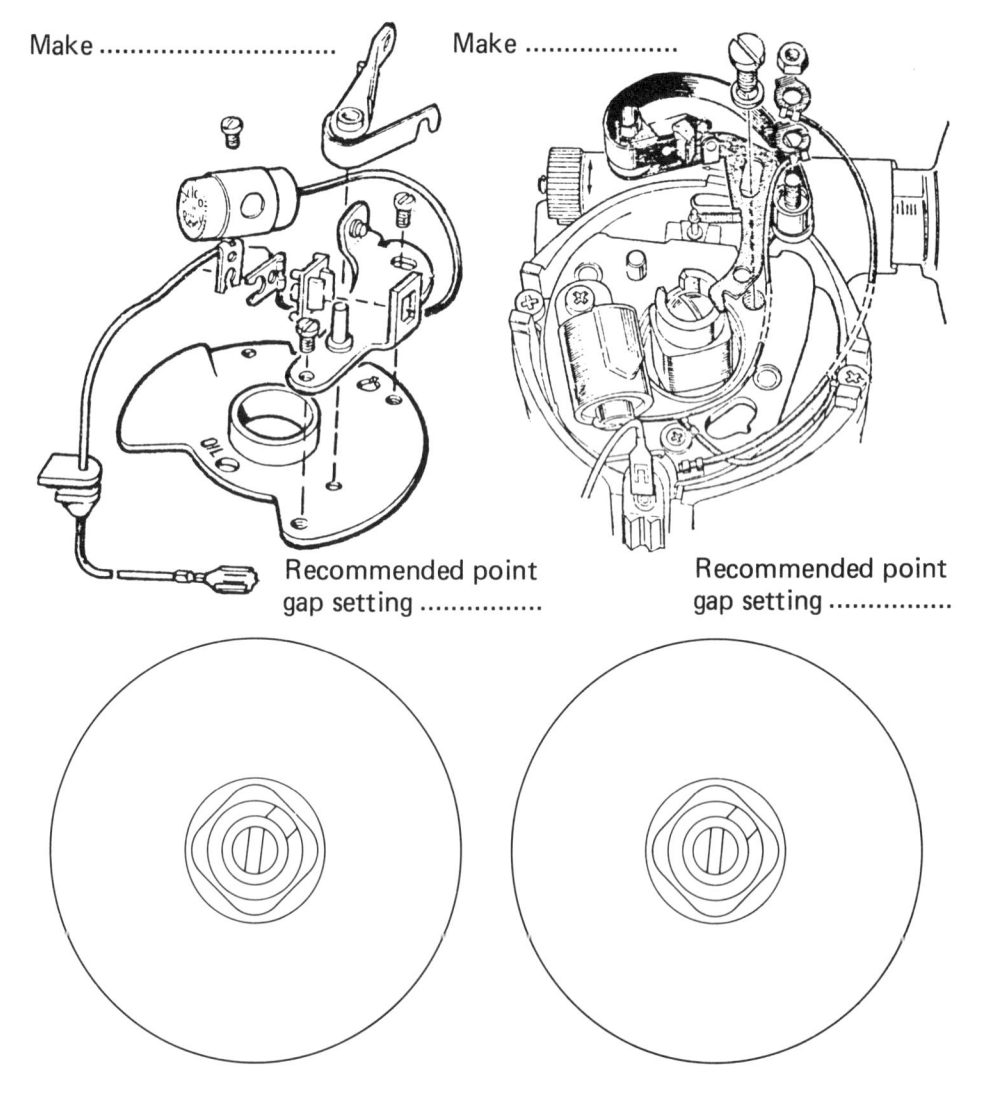

Make ....

Make ....

Recommended point gap setting ....

Recommended point gap setting ....

DWELL ANGLE

What is meant by 'dwell angle'?

....

....

....

....

....

....

Complete the drawing to show dwell angle.

Distributor cam

What effect would the following incorrect dwell angle have on coil performance and therefore the running of the engine?

Excess dwell — ....

....

Insufficient dwell — ....

....

Quote different manufacturers' specifications for four-cylinder distributors fitted to various vehicles.

Distributor make	Vehicle		Dwell angle	C.B. gap setting
	make	model		

CAPACITORS

What is the function of the capacitor?

....

....

....

Electrically speaking, where in the ignition circuit is the capacitor fitted?

....

....

....

CONSTRUCTION

The capacitor consists of two sheets of tin foil or metalised paper which are rolled together but electrically insulated from one another. Sketch details of the construction below.

State three capacitor faults and their symptoms, that may cause ignition troubles.

1. Fault ....

   Symptom ....

2. Fault ....

   Symptom ....

3. Fault ....

   Symptom ....

ADVANCE AND RETARD MECHANISMS

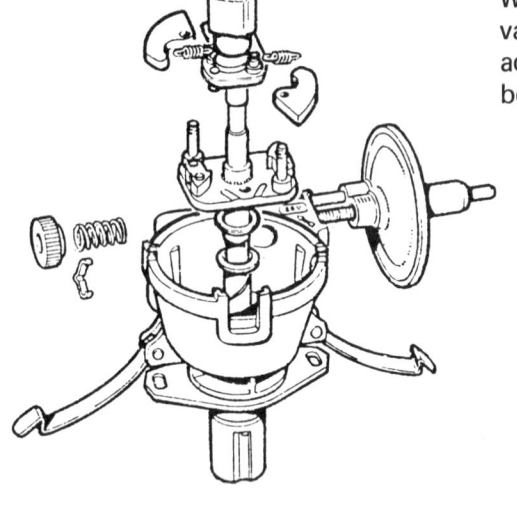

When the engine speed or load is varied it is usual to automatically adjust the ignition timing. This may be done in two ways:

(i) ....

....

....

(ii) ....

....

....

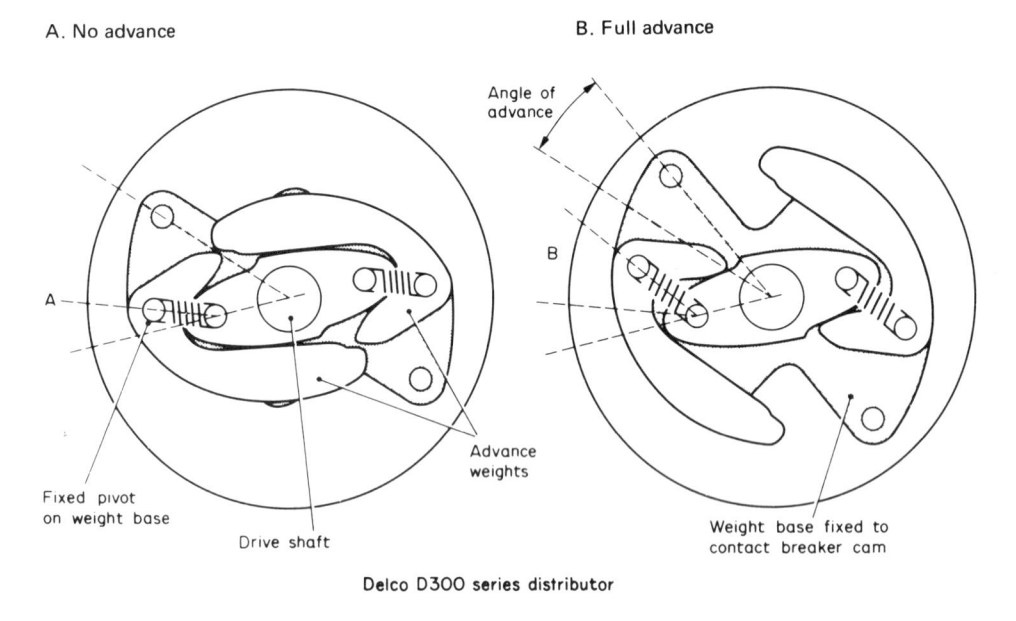

A. No advance

B. Full advance

Angle of advance

B

A

Fixed pivot on weight base

Drive shaft

Advance weights

Weight base fixed to contact breaker cam

Delco D300 series distributor

CENTRIFUGAL ADVANCE

The type shown above has pivoted weights mounted below the contact breaker base plate.

The type shown opposite has the advance weights mounted above it.

Examine similar distributors and other types, explain their principle of operation and sketch an alternative weight mechanism in the spaces opposite.

....
....
....
....
....
....

158

VACUUM ADVANCE

Complete the drawings to show the operation of the vacuum unit and its connection to the distributor base plate and carburettor.

Name the various parts.

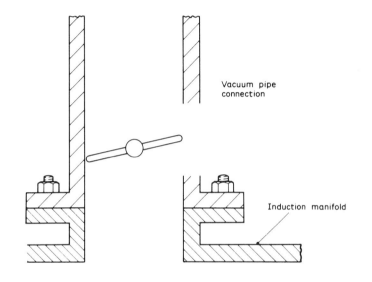

Vacuum pipe
connection

Induction manifold

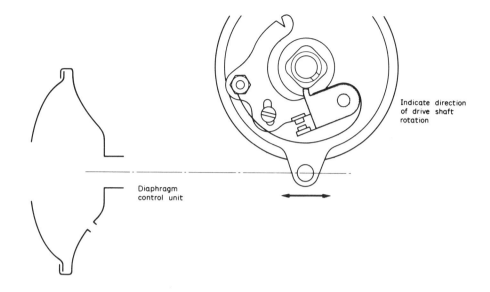

Indicate direction
of drive shaft
rotation

Diaphragm
control unit

Explain how this type of advance and retard mechanism operates.

....
....
....
....
....
....
....
....
....

On the graph below show typical advance curves for vacuum and mechanical mechanisms.

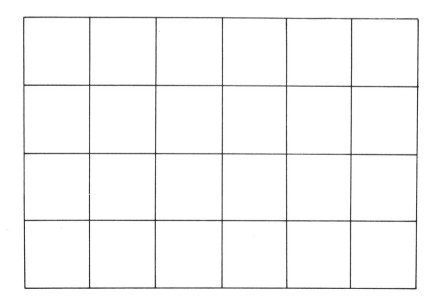

SPARK PLUGS

The high-voltage energy produced by the coil is dissipated in the form of a spark across the spark plug electrodes. This ignites the petrol/air mixture in the combustion chamber.

The spark plug must operate efficiently under widely varying conditions of pressure and temperature and must be designed to suit the type of engine to which it is fitted.

State the importance of the following variations in plug design:

Plug reach — long or short ....
....
....

Thread diameter ....
....
....

Methods of sealing plug seat ....
....
....

Earth electrode design ....
....
....

Most modern vehicles use plugs having an extended nose (as shown).
What are the advantages of using such a plug?

....
....
....

On the sectioned view of the spark plug shown, the arrows indicate certain special features. Name and comment on these features.

....
....
....
....
....
....
....
....
....
....
....
....
....
....
....
....
....
....
....
....
....
....

State a typical plug gap setting. ....

At what recommended mileage should plugs be replaced? ....

VARIATION IN PLUG DESIGN

HEAT RANGE

For any particular application spark plugs must be selected that do not foul at slow speeds or get red hot at high speeds.

....
....
....
....
....
....
....
....

The graph below shows how the temperature of the insulator tip varies and how it is affected if the incorrect plug is fitted.

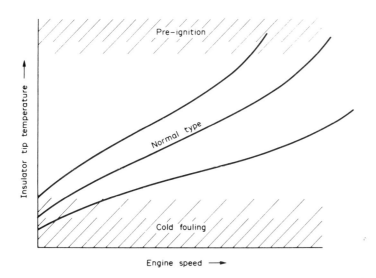

Complete the diagrams below to show the difference in construction between a hot and a cold spark plug, indicate with arrows the heat-flow path.

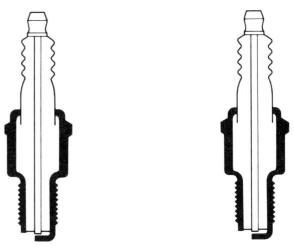

Examine a manufacturer's chart giving spark plug specifications.

Select a range of plugs and complete the table below.

Make of spark plugs considered ....

Plug number, place coldest plug first	A typical vehicle for which plug is recommended	Correct gap setting

ENGINE RUNNING CONDITIONS

If possible examine a manufacturer's spark plug condition colour chart, together with uncleaned spark plugs removed from various engines.

State the plugs likely appearance when the engine's running condition is said to be:

Normal	....
Mixture too weak	....
Mixture too rich	....
Engine overheated	....

What are the effects of running an engine with the plug electrodes set,

(i) too wide ....

....

(ii) too close ....

....

Why does filing the plug electrodes lower the voltage required to fire the plug? ....

....

HIGH-TENSION CIRCUIT

The high-tension circuit consists of the coil distributor cap, rotor, plugs and plug leads.

....

....

....

....

Two ways in which the high-tension lead may be held in the distributor cap are:

(i) .... (ii) ....

State what material the distributor cap is made from and give reasons for its use.

....

....

....

What electrical property does the rotor contact brush that is fitted into the cap often incorporate?

....

On certain high-speed engines the maximum speed is controlled by a rotor cut-out. How does this operate?

....

....

....

....

....

HIGH-TENSION LEADS

Two types of leads are commonly used.

1. The centre core made of stranded copper wire.

2. The centre core made of stranded and woven rayon or silk and impregnated with graphite.

What are their relative advantages?

1. Copper core leads ....
....
....

2. Graphite leads ....
....
....

Explain how the terminal connections are made on the graphite impregnated leads.

....
....
....

Show two types of high-tension coil connections.

SUPPRESSORS

Whenever an electrical spark occurs waves of electrical energy are radiated out from the spark source. This energy causes external electrical interference which by law must be suppressed. (Wireless Telegraphy — Control of Interference from Ignition Apparatus — Regulations 1952.)

Resistors having a value of between 5000 and 25 000 ohms are required, these may be incorporated in at least four components.

....
....
....

EFFECTS OF DAMPNESS ON IGNITION SYSTEMS

Why does dampness affecting the ignition system make cars difficult to start or cause engines to misfire and even stop?

....
....
....

Explain the effect 'flashover' (spark tracking) has on distributor caps.

....
....
....

State two methods by which a damp ignition system could be dried out.

1. ....

2. ....

ELECTRONIC IGNITION SYSTEMS

Electronic ignition systems use transistors as high-speed switches in the circuit, carrying the heavy current normally handled by the contact breaker points.

The system may be of an inductive — which is the most popular — or a capacitive discharge design.

Either design can be subdivided into two types:

1. ....

2. ....

The simple systems are the T.A.C. (transistor assisted contacts) types.

....

....

....

....

With the contactless systems (see diagrams) the contact breaker is removed, and replaced with a trigger system that is either optical or magnetic. The only moving parts in the distributor that may then wear are the advance—retard mechanisms.
How are the transistors triggered in the optical system?

....

....

....

Name the indicated parts

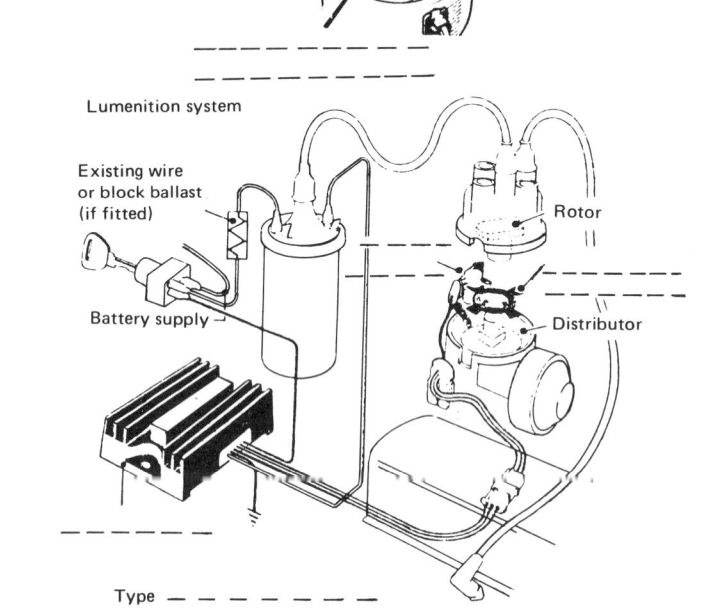

Mobelec system

Yellow

Green

Black

Red

To ignition switch

Coil

Distributor

Type

Lumenition system

Existing wire
or block ballast
(if fitted)

Rotor

Battery supply

Distributor

Type — — — — — —

How are the transistors triggered in the magnetic system?

....

....

....

....

....

....

In the capacitive discharge systems the 12 V input is boosted to 400 V and used to charge a capacitor. When the system is triggered it is released into the primary coil allowing the secondary circuit to build up to voltages of around 40 000 V.

What advantage does this create?

....

....

....

....

State the advantages of electronic ignition when compared with the conventional system.

....

....

....

....

....

....

IGNITION TIMING

In order to ensure that the pressure build-up in the combustion chamber occurs just as the piston passes t.d.c. the spark must be timed to occur approximately 5 to 10° before t.d.c. when the engine is turning at idling speeds.

As the engine speed is increased, the ignition timing (point when ignition occurs) must be automatically advanced.

....

....

....

....

State the effects of incorrect ignition timing.

(i) Over-advanced ....

....

(ii) Retarded ....

....

Describe the method of statically checking the ignition timing, using a test lamp.

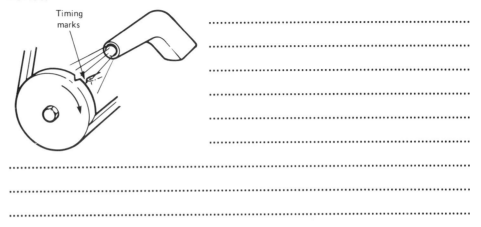

....

....

....

....

Describe a method of checking the ignition timing using a stroboscopic device.

....

....

....

....

....

....

....

INVESTIGATIONS

1. Run an engine with correct ignition timing at a speed equivalent to about 30 m.p.h. in top gear; then successively over-advance and over-retard the engine and note the effect on its running.

....

....

....

....

2. Check the operation of mechanical and vacuum advance mechanisms.

(a) Accelerate engine and note maximum amount of advance. This is both the mechanical and vacuum advance.

  Actual reading ....   Manufacturer's spec. ....

(b) Disconnect the vacuum pipe and accelerate engine. This will give the mechanical advance only.

  Actual reading ....   Manufacturer's spec. ....

RESISTANCE VALUES

In the ignition circuit the most common items associated with electrical resistance are the ignition leads and the amount of resistance required to provide satisfactory suppresion.

What creates the electrical interference which by law must be suppressed?

....

State a typical suppressor lead resistance valve. ....

INVESTIGATION

Using an ohm-meter, measure the resistance of various items of equipment in the ignition system.

Item	Resis value Ω	Item	Resis value Ω	Item	Resis value Ω
Ignition lead		Coil — Low tension circuit		Ballast resistor	
Brush in distribu-tor cap		Coil — High tension circuit		Spark plug	

COIL OUTPUT

The coil's secondary voltage builds up until it creates a spark at the plugs. When these are in good condition the voltage needed may be as low as........... but with constant use and engine faults can rise to.....................................
The coil usually has a maximum output capacity of.....................................
With some transistorised coils it can be as high as.......................................

State what factors influence the coil's output.

....

....

....

....

PRIMARY CIRCUIT CONNECTIONS

Most coils are wound and connected to produce a negative spark at the spark plugs.

Why is this? ....

....

....

....

....

The effects of wiring the coil to give incorrect polarity would be:

....

....

....

....

Show the coil top markings when the primary circuit is wired correctly.

From ignition switch   To distributor

POSITIVE EARTH

From ignition   To distributor

NEGATIVE EARTH

HIGH-TENSION VOLTAGE

The spark plug firing voltage is affected by variables in the engine cylinder and in the plug itself.

Factors which would raise the required voltage and therefore lower the plugs efficiency are:

....

....

....

....

INVESTIGATION

Using suitable test equipment, observe the variation in voltage required to fire a spark plug or the pressure required to make the spark misfire when the spark plug gap is varied.

Equipment used ....

Type of spark plug ....

Plug gap setting					
Plug misfiring occurs	Voltage				
	Pressure				

EFFECTS OF PRIMARY CIRCUIT VOLTAGE

The quality of the induced secondary voltage (spark) depends to a very great extent on the rate at which the magnetic field collapses. This being the function of the primary circuit in particular of the capacitor (condenser).

State factors that affect the primary–secondary voltage relationship.

....

....

....

....

....

....

Explain, with the aid of the graph below, the function of the capacitor in producing a high voltage at the coil, sufficient to cause a spark at the plug.

....

....

....

INVESTIGATION

Examine the effects of producing a spark with and without a capacitor.

....

....

RELATIONSHIP OF DWELL ANGLE TO IGNITION TIMING INVESTIGATION

To determine what the effects of varying the dwell angle has on ignition timing.

Using a dwell meter, with a distributor tester or running engine and timing light, take dwell and ignition timing readings at three contact breaker gap settings.

Method used ....

....

Distributor make .... Specified dwell angle ....

Adjustment	Contact breaker gap	Dwell angle	Ignition timing
Correct setting			
Too wide			
Too narrow			

What is the relationship between the dwell angle, contact breaker gap, and ignition timing?

....

....

VOLTAGE INDUCTION INVESTIGATION

Using a simple rig, show how an e.m.f. can be induced into a circuit by varying the current flow in a neighbouring but separate circuit.

....

....

....

....

Show sketch of equipment used, naming parts.

Describe how an e.m.f. is induced by mutual induction.

....

....

....

....

Use an ignition coil in a similar manner to produce a spark.

Constructionally in what way does the coil differ from the apparatus above?

....

....

....

....

State typical coil ignition operating voltages.

Low-tension voltage	Acceptable high-tension voltage for normal engine running	Maximum high-tension voltage

EMISSION CONTROL

APPLIED STUDIES

POLLUTANTS CREATED BY THE MOTOR VEHICLE

The creation of some pollution is inevitable when power is obtained by internal combustion using a hydrocarbon fuel; and when an engine is not running in tune an excess of harmful emissions are passed into the atmosphere.

Name the three basic areas that may emit pollutants into the atmosphere.

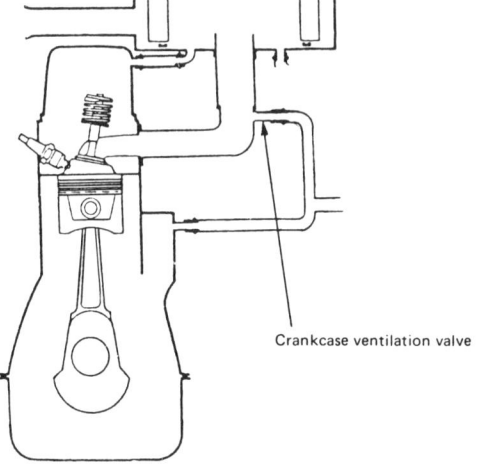

....

....

....

Name the pollutants produced by an engine and state why they are there.

....

....

....

....

....

CONTROL OF EVAPORATIVE AND CRANKCASE EMISSION

Having recognised the need to control pollution, it is relatively easy to control crankcase emissions (covered in detail on p.29 and p.30) and fuel loss through evaporation.

How are these emissions controlled? ....

....

....

Complete the drawing below to show the layout of a typical sealed fuel ventilation system. Indicate also the direction of air flow through the engine's crankcase emissions arrangement.

Crankcase ventilation valve

Positive sealing fuel cap

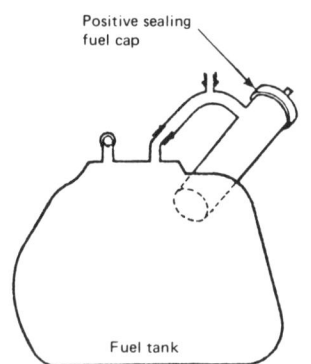

Fuel tank

....

....

If a greater control of fuel evaporation is required (as in USA) a charcoal canister filter is positioned in the vapour line system.

....

....

CONTROLLING EXHAUST POLLUTION

European emission regulations (ECE 15) are much less stringent than US legislation.

Quote typical European and US maximum emission values.

Exhaust emission g/mile	CO	HC	NO_x
European			
United States			

In the USA these values must be maintained for ....

To achieve the European values it has only therefore been necessary to improve engine-running conditions to ensure that the fuel is burnt most efficiently in the combustion chamber.

What other advantage does this create.

....

....

Describe examples of how engines and components have been improved to reduce harmful emissions.

Fuel ....

....

Ignition ....

....

....

....

Carburettor ....

....

....

....

Combustion ....

THERMOSTATICALLY CONTROLLED AIR INLET VALVE

Used to achieve a quick warm-up. Show valve's position for each of the stated titles.

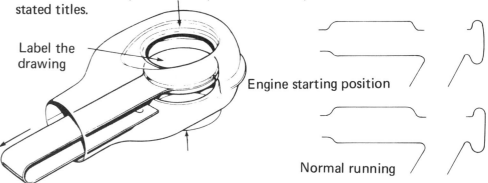

Label the drawing

Engine starting position

Normal running

In order to achieve much lower emission from the combustion area the Honda CVCC engine uses a pre-combustion chamber and an overall weak mixture.

....

....

....

....

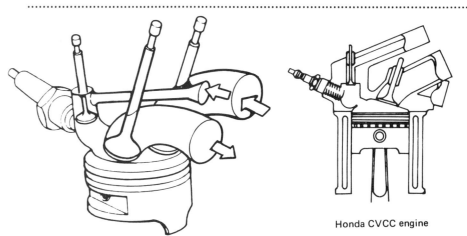

Honda CVCC engine

171

When more stringent emission controls are required (e.g. cars to be exported) 'add on' devices become necessary to treat and rid the exhaust fumes of their harmful emissions after they leave the combustion chamber.

Two examples of such devices are shown on this page.

AIR INJECTION SYSTEM

The method shown below blows fresh air into the exhaust manifold at points opposite each exhaust valve. The extra air helping to burn any CO or HC left in the waste gas.

Name the parts indicated.

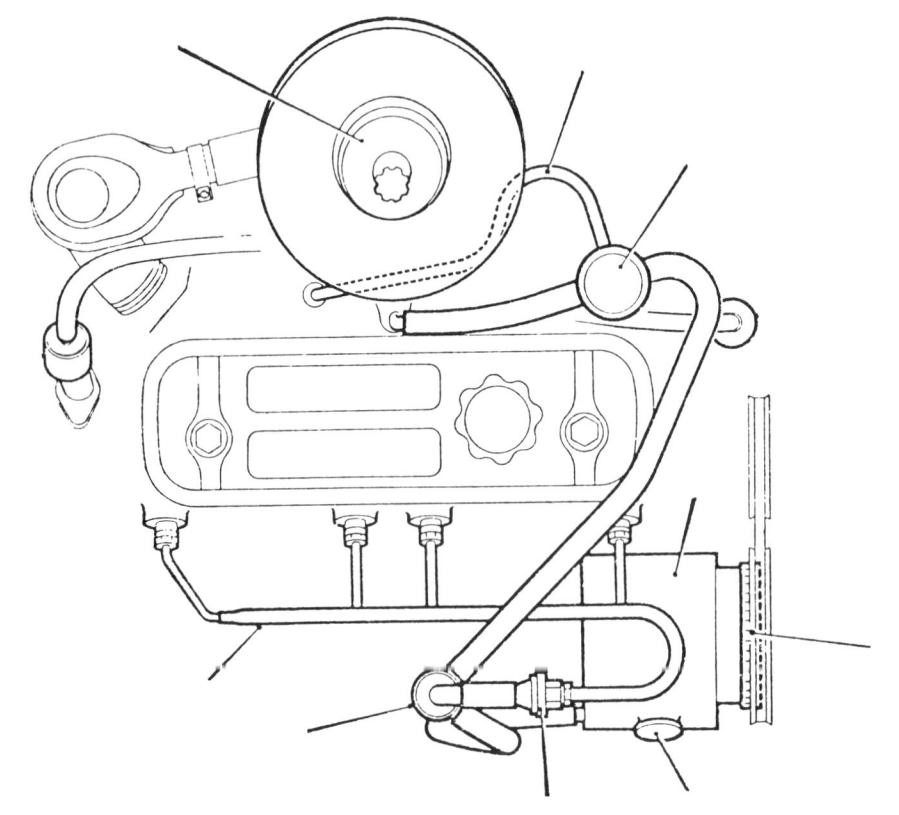

What is the purpose of the check valve? ....

....

What is the purpose of the gulp valve? ....

....

CATALYST CONVERTOR SYSTEM

An alternative method is to use a catalytic convertor. This contains a bed of platinum/palladinum coated pellets which absorb the HC and CO by a chemical process and convert them to harmless H_2O (...............), and CO_2 (...............)

Show the exhaust gas flow through the catalyst box below and name the main parts.

Exhaust gas in

What are the disadvantages of this system?

....

....

....

EXHAUST GAS RECIRCULATION (EGR)

A further method commonly used is to allow about 10% of the exhaust gas to recirculate back into the carburettor.

Its object being to ....

....

172

POLLUTION FROM COMPRESSION IGNITION ENGINES

From the harmful emission point of view, if the air:fuel ratio is maintained around 22:1 emissions of CO, NO_x and HC are kept within acceptable limits. However, any deviation from correct tune, will create an excess of unburnt hydrocarbons which will be ejected in the form of smoke.

COMMON CAUSES OF SMOKE

Smoke may be described in three forms as listed below.
State the basic causes of these colours giving reasons why they have occurred.

Smoke colour	Cause	Possible faults
Black		
Blue		
White		

METHOD OF MEASURING SMOKE DENSITY

A smoke meter must be used if the density of smoke is to be measured accurately.

Diagram shows a Hartridge smoke meter mounted on a portable cabinet connected to a vehicle being tested on a chassis dynamometer; there it can measure accurately the amount of smoke produced under different conditions of load, speed and power output.

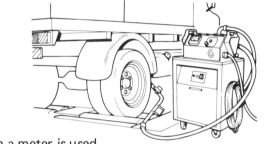

Describe how such a meter is used.

....

....

....

....

....

....

....

What are the maximum permissible smoke levels (BS AU 141a)?

....

....

173

APPLIED STUDIES

EFFECTS OF MIXTURE STRENGTH ON EXHAUST EMISSIONS

Examine the graphs and comment on how varying the mixture strength can alter the exhaust emissions from spark-ignition engines.

....

....

....

....

....

....

....

....

....

....

....

....

....

....

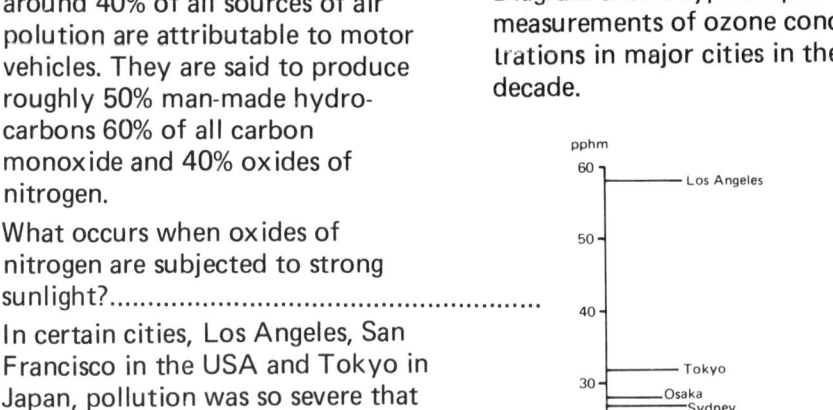

EFFECTS OF POLLUTION ON THE ENVIRONMENT

Rightly or wrongly it is said that around 40% of all sources of air pollution are attributable to motor vehicles. They are said to produce roughly 50% man-made hydro-carbons 60% of all carbon monoxide and 40% oxides of nitrogen.

What occurs when oxides of nitrogen are subjected to strong sunlight?.......................................

In certain cities, Los Angeles, San Francisco in the USA and Tokyo in Japan, pollution was so severe that stringent anti-pollution laws were brought into effect.

In Western Europe climatic conditions are not conducive to smog and therefore it is unlikely that this area will ever need such strong controls.

Diagram shows typical spot measurements of ozone concentrations in major cities in the last decade.

Threshold limit value is one defined by the DoE as acceptable

State the effects of the main pollutants produced by motor vehicles.

Pollutant	Symbol	Effects
Carbon monoxide		
Hydrocarbon		
Oxides of nitrogen		
Carbon		
Lead		

174

COOLING

APPLIED STUDIES

COOLING SYSTEM

The principle of the heat engine which derives its power from burning fuels and the resultant expansion of gases necessitates the use of some type of cooling system. During combustion the temperature in the cylinder can momentarily be as high as 1800°C. Even when the gases expand and the temperature falls, it may still be higher than the melting-point of aluminium.

What problems may occur if this heat is not dissipated at the correct rate?

....
....
....
....

There are two main types of cooling system:

1. .... 2. ....

List the relative merits of both these systems.

....
....
....
....
....
....
....
....
....
....
....
....

LIQUID (WATER)-COOLING SYSTEM

This basically consists of water jackets surrounding the cylinders with provision for the heated water to pass into a radiator and cooled water from the radiator to flow back into the cylinder block.

Using arrows indicate the flow of water through the system.

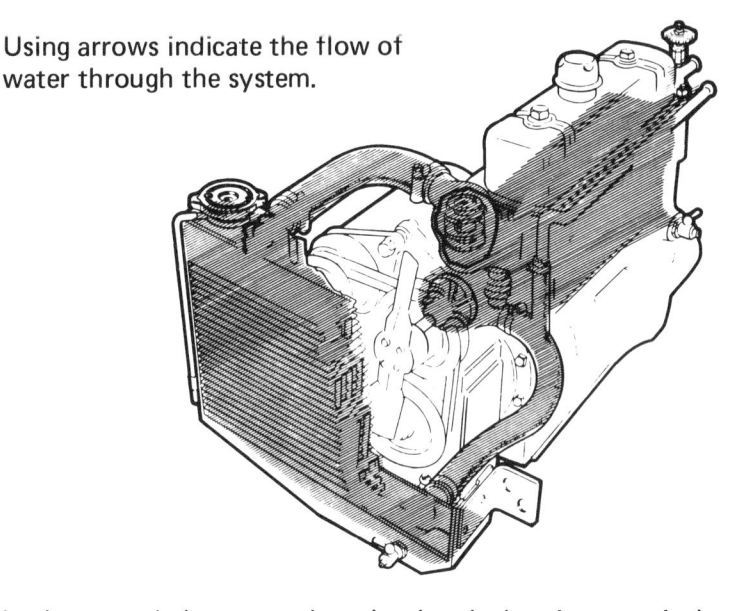

In the space below, complete the sketch showing a typical water cooling system. Include the water pump, thermostat and pressure cap

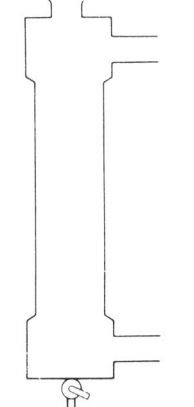

AIR-COOLING SYSTEM

With a few notable exceptions this system is not very popular for multi-cylinder engines as difficulties are encountered trying to cool equally all the cylinders and maintain a constant temperature.

How is the air flow normally controlled?

....
....
....
....
....
....

Show the direction of air flow through the FIAT engine shown below, and name the parts indicated.

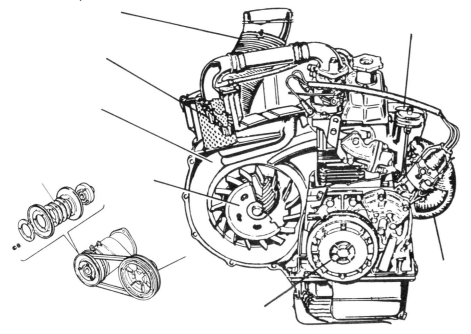

Indicate the air flow through the horizontally opposed engine shown below, and name important parts.

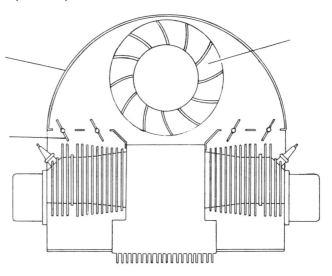

What are the advantages of horizontally opposed cylinders, with regard to air cooling?

....
....

Why is it particularly important on an air-cooled engine that fan-belt tension is correct?

....
....
....
....

The air fan is commonly driven from the rear of the generator and fan-belt tensioning is made by means of a split pulley.

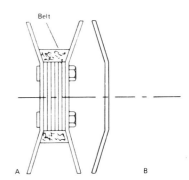

Complete sketch of pulley B to show how the belt may be tensioned.

WATER-COOLED PRESSURISED SYSTEM

Water is a better cooling medium than air; it has a high specific heat and it conducts heat more efficiently.

Unfortunately for water-cooling systems the engine gives its best thermal efficiency when the cooling water is close to 100°C (i.e. normal boiling-point). One consequence is that systems are normally pressurised.

What advantages are gained by pressurising the system?

....

....

....

....

....

THERMOSTATS (see also p. 189)

A thermostat is a disc valve which when the water is cold remains closed, and as the engine reaches its normal running temperature expansion occurs within the thermostat (i.e. valve control unit) and the disc valve opens.

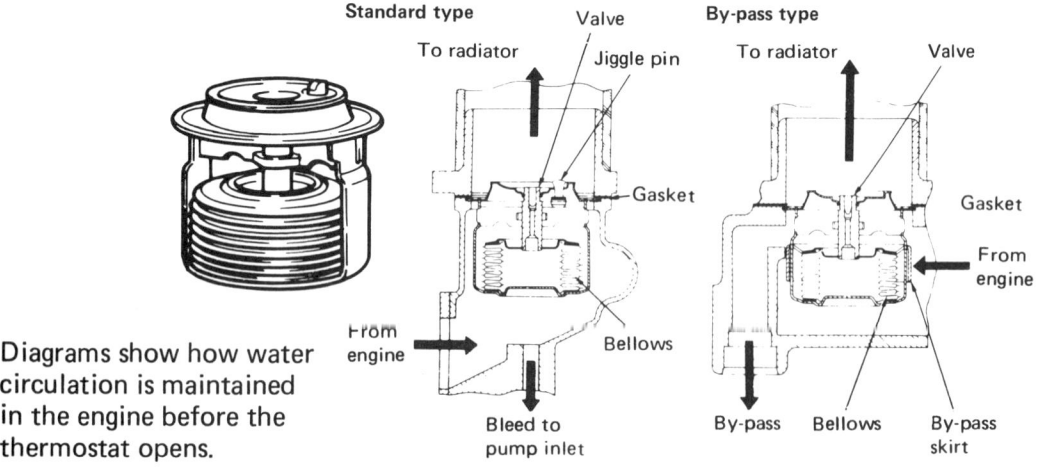

Diagrams show how water circulation is maintained in the engine before the thermostat opens.

BELLOWS TYPE

This type (now rather old-fashioned) uses a bellows action to open the valve. The lower end of the bellows is fitted to a frame whilst fastened to the top and is the valve.

Describe its basic operation.

....

....

....

....

WAX ELEMENT TYPE

This type employs a special wax contained in a strong steel cylinder into which passes the thrust pin. This is surrounded by a rubber sleeve which also seals the upper end of the cylinder.

The wax element type thermostat is shown fully closed on the left. Complete the right-hand sketch showing the element in the open position. Add the names of the important parts.

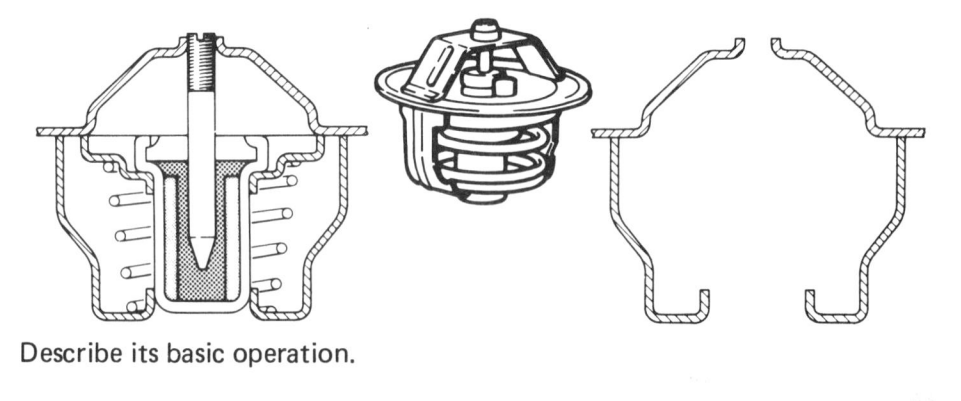

Describe its basic operation.

....

....

WATER PUMPS

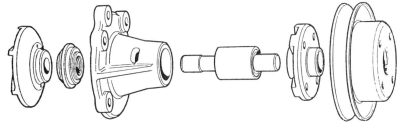

Modern water pumps, properly called impellers, are usually bolted to the front of the cylinder block and belt-driven from the crankshaft.

....

....

....

....

Shown below is a sectional view of the casing of a typical water pump body. Sketch in the missing parts, and name them.

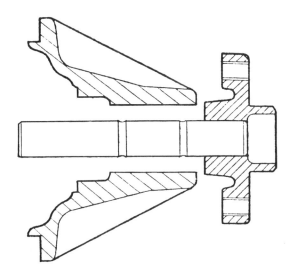

Explain the provision normally made to circulate the water expelled from the water pump when the thermostat is closed.

....

....

....

PRESSURE CAP

This consists of a spring-loaded valve which resists the pressure of the expanding coolant, air and steam, in the header tank unit.

How is the pressure controlled?

....

....

....

Explain the purpose of the small valve fitted in the centre of the pressure cap.

....

....

Complete the sketch below by adding the missing half of the pressure cap in section. Name the parts.

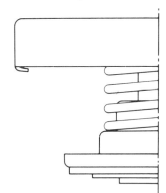

ALTERNATIVE COOLING SYSTEM LAYOUT

It is becoming more common with modern designs of vehicles to place the radiator, pump, thermostat and fan in positions other than directly in front of the engine.

Below is shown one such arrangement. Name the parts indicated and show the direction of water flow when:

(1) The engine is cold (during rapid warm-up).

(2) The engine is at normal running temperature.

Give reasons for the alternative positioning of water pumps.

....

....

....

....

....

SEALED COOLING SYSTEMS

A disadvantage of the ordinary cooling system is that small losses of coolant occur through the radiator overflow pipe. If the level of water in the header tank is not frequently checked it is possible for the water level to fall sufficiently to prevent circulation. A method of overcoming this is to simply immerse the lower end of the overflow pipe in coolant contained in an expansion chamber.

....

....

....

....

Why is a sealed system considered essential on some heavy vehicles and P.S.V.s.

....

....

List the advantages and therefore reasons for use of a sealed cooling system.

....

....

List the disadvantages of a sealed cooling system.

....

....

RADIATORS

The purpose of the radiator is to provide a large cooling area for the water and expose it to the air stream. A reservoir for the water is also included in the construction, this is known as the header tank and is normally made from thin steel or brass sheet. The header tank is connected to the bottom tank by numerous brass or copper tubes surrounded by cooling 'fins' and this assembly is known as the matrix, block, stack or core.

MATRIX CONSTRUCTION-TUBE TYPE

This consists of thin, almost flat or oval copper or brass tubes arranged in rows.

....

....

....

....

....

Name the main parts

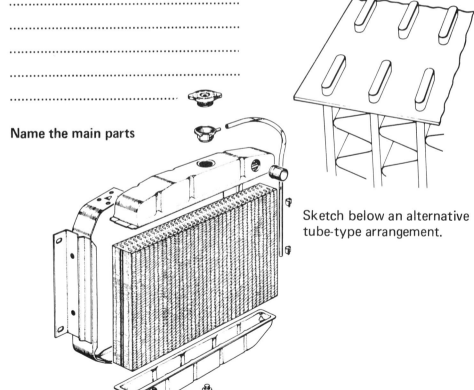

Sketch below an alternative tube-type arrangement.

FILM TYPE

....

....

....

....

....

....

....

Examine a film-type matrix construction and in the space provided below make a sketch similar to the previous two.

COMMERCIAL VEHICLE TUBE TYPE

The top and bottom tanks are often aluminium alloy castings whilst the gilled copper tubes are of circular cross-section. These tubes are normally fitted separately and a water tight joint is obtained in some cases by using rubber seals. In other instances the copper tubes are soldered into upper and lower plates which bolt on to the top and bottom tanks.

Complete the sketch below by adding details of how the tube is connected to the base plate of the top tank.

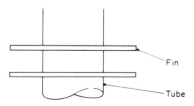

Fin

Tube

CROSS-FLOW RADIATORS

In the conventional radiator the coolant flows ....
through the core from .... to ....

In cross-flow radiators the coolant flows ....
through the core from the .... of one side tank across to the
.... of the other side tank.

Indicate the direction of water flow through the radiator shown.

Why is it considered necessary to fit such a radiator in preference to the
vertical flow type?

....

....

....

What are the cross-flow radiator's basic disadvantages?

....

....

....

....

With most cross-flow systems it is necessary to fit a remote header or
expansion tank as shown below.

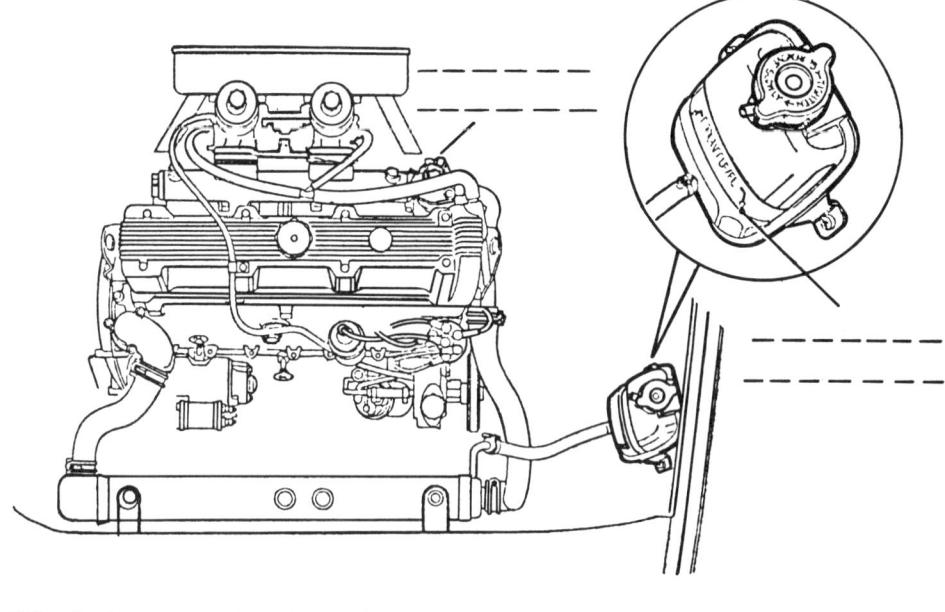

Why is the remote header tank necessary?

....

....

....

Describe the coolant filling procedure for the layout shown.

....

....

....

....

....

....

FANS

Most cooling systems (water or air) are fitted with fans. In the simplest arrangement the fan is permanently driven from the crankshaft via the fan belt.

The fan's function is:...

....

Why is the type of fan described above becoming much less popular on modern vehicles:

....

....

AUTOMATICALLY CONTROLLED FAN

These fans will adjust the air speed to cooling requirements by either: altering the fan-blade pitch, switching on when required or varying the actual speed of the fan.

List the advantages of automatically controlled fans.............................

....

....

VARIABLE-PITCH FAN

Probably the simplest type is the version where the pitch of the blade is at its coarsest when stationary and as centrifugal force increases the blades begin to feather...

A more sophisticated design is where a coiled bi-metal strip is mounted in the boss of the fan unit. As the air temperature flowing through the radiator increases the bi-metal strip expands. The inner ends of the fan blades are attached to the strip and pass through off-set elongated holes, which alters the pitch of the blades. The angle of pitch can alter from 0° to approximately 45°.

....

FREE-WHEELING FAN

This type is operated electrically and when switched on the electro magnet in the drive pulley magnetises and locks the free-wheeling fan and armature.

Name the arrowed parts.

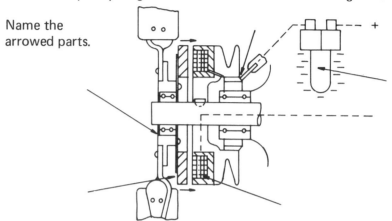

What operates the thermostatically controlled switch?..............................

Where would it be positioned?...

....

ELECTRICALLY DRIVEN FAN

The construction of this type includes an electric motor complete with fan; a temperature-sensitive control unit; a warning light; and an override control.

Describe the operation.

....

....

....

....

....

....

VISCOUS COUPLINGS TORQUE-LIMITING FAN

These units operate as shear-type fluid couplings.

....
....
....
....
....
....

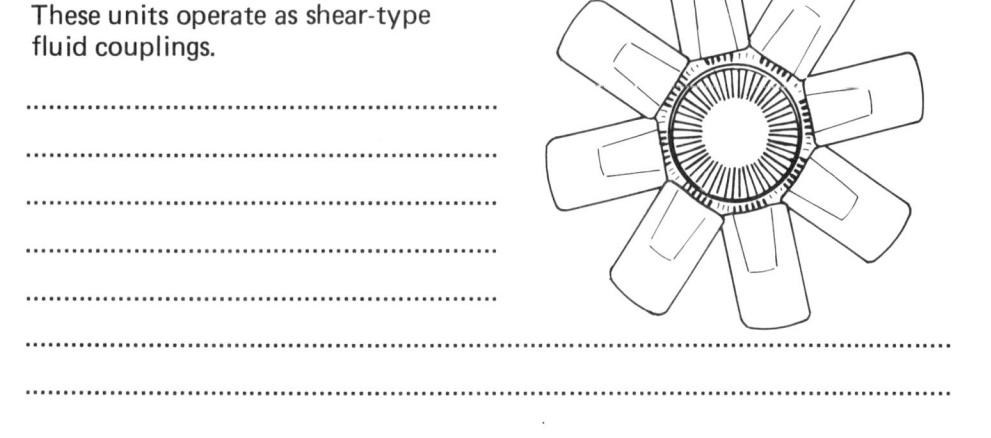

AIR-TEMPERATURE-SENSITIVE FAN

These also operate on the viscous shear principle. The unit is divided into two chambers: one contains the driving disc, the other a reservoir.

....
....
....
....
....
....
....

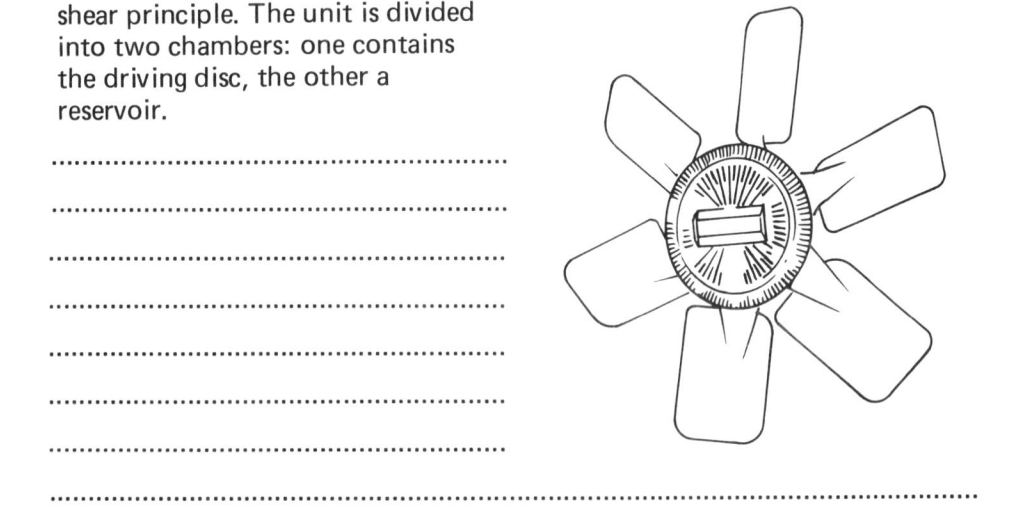

Name the important parts.

Show on the graphs the typical performance of these drives compared with normal belt-drive.

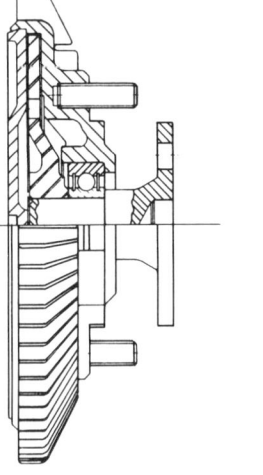

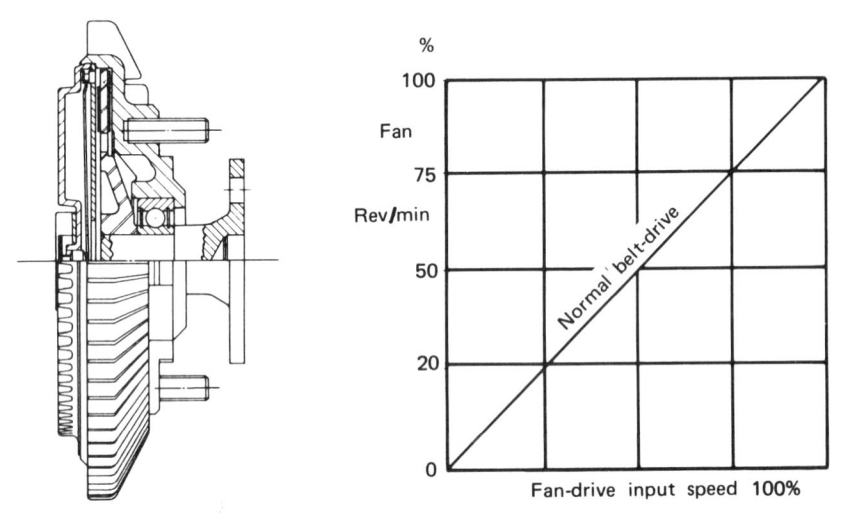

Torque-limiting design.

Air-temperature-sensitive design.

WATER HEATERS

To give adequate comfort to the occupants of a vehicle in varying weather conditions, all modern vehicles are equipped with some form of heating and ventilation system. The heat source, on vehicles fitted with water-cooled engines, is usually the hot water from the engine. Heater units which utilise the engine cooling water are normally situated on the bulkhead.

....
....
....
....
....
....
....
....
....

A heater system, not shown on this page, is known as a 'recirculatory' type.

How does this differ from the conventional type?

....
....
....
....
....

Drawing shows a fresh-air cab heater with its operational flaps in the off position.

Show the flaps positioned to give:
(i) Maximum heat to interior and screen.
(ii) Warm air to screen only.

Modern heating and ventilation layout.
Using arrows, show the flow of cool and heated air.

What does the drawing below show?

....
....
....

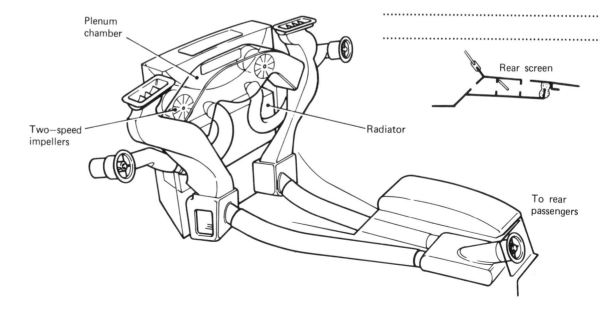

Plenum chamber

Two–speed impellers

Radiator

Rear screen

To rear passengers

185

COMBUSTION HEATER SYSTEM

These are interior heaters which work independently of the engine cooling system.

Below is shown a schematic layout of such a system.

Indicate the air flows and describe its operation.

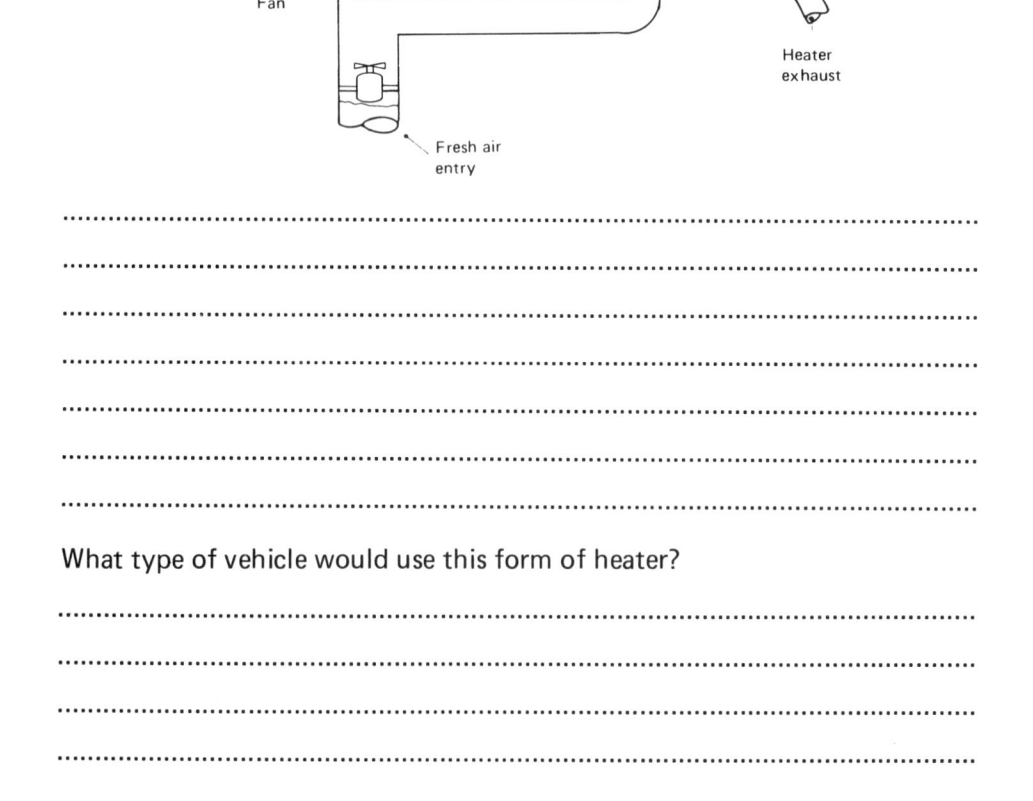

....

....

....

....

....

....

....

What type of vehicle would use this form of heater?

....

....

....

....

COOLING SYSTEM FAULTS

List the possible causes of the cooling system faults listed below.

Fault	Cause
Engine overheating	
Engine overcooling	
Corrosion	
External leakage	
Internal leakage	

186

APPLIED STUDIES

ANTIFREEZE

One of the disadvantages of using water as a coolant is the problem of freezing. Water possesses its greatest density at 4°C and when its temperature is lowered (or raised) it begins to expand. The expansion rate becomes a serious problem as freezing occurs. Great force can be exerted by ice and can result in cracked cylinder blocks and heads. Fortunately the freezing-point of water can be lowered considerably by the addition of certain liquids. An example of such a substance is ethylene glycol.

What is the recommended anti-
freeze mixture in...........................?

....

....

Using the following values, plot a graph to show the liquid to ice freezing relationship of the coolant mix.

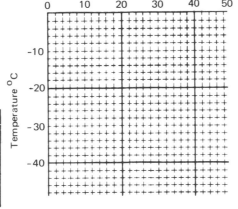

Ethylene glycol (%) in coolant

% Ethylene glycol	25	33.3	40	50
Coolant begins to mush	−12	−20	−24	−36
Coolant becomes ice	−26	−35	41.5	−48

The graph shows two of the most important properties in an ethylene-glycol-based antifreeze.

....

What other properties must an antifreeze possess?

....

....

What are the disadvantages of using a methanol-based antifreeze?

....

EFFECTS OF CORROSION – ACID AND ALKALIS

What are the main causes of corrosion in the cooling system?

....

....

....

....

....

....

How does continual heating affect the mineral deposits?

....

....

....

....

Radiator materials

What materials are used in the construction of radiators and what are the reasons for their choice?

....

....

....

....

....

HEAT LOSSES

Heat losses occur when heat transfers from hot substances to cold substances.

....

If a small, medium and large engine were all operating at the same running temperature, which, when switched off, would cool down the quickest?

....

Heat energy is measured in .... It can be directly converted or expressed as mechanical work units where:

1 Newton metre =

The amount of heat contained in 1 kg of substance is known as its

....

The heat required to raise 1 kg of water 1°C is...
∴ The specific heat capacity of water is...

Different materials accept (or lose) heat at different rates and therefore for a similar mass they will increase (or decrease) their temperature at different rates.

When heated at the same rate 1 kg of oil will increase its temperature at a
.... rate than 1 kg of water.

Substance	Specific heat capacity kJ/kg °C	Substance	Specific heat capacity kJ/kg °C
Water		Steel	
Lubricating oil		Brass	
Aluminium		Lead	

List below the information it would be necessary to know, to be able to calculate the quantity of heat transferred from one substance to another.

....

Problems   Note! 1 litre of water has a mass of ....

Heat lost or gained/s = mass flow/s × specific heat capacity of substance × temperature change

1. A pump circulates 150 litres of water through a cooling system in 2 min. The temperature at the top of the radiator is 90°C and at the bottom 70°C. Calculate the heat energy radiated per second.

2(a) A cooling system contains 15 kg of water. Calculate the quantity of heat gained by the water if its temperature rises from 12°C to 88°C on starting.

(b). What heat is lost/s during cooling if the flow rate is 2 l/s and the temperature at the bottom of the radiator is 53°C?

EFFECTS OF EXTREMES OF TEMPERATURE

....

....

....

....

....

....

....

....

CHANGES OF STATE

All substances exist in one of three states: solid, gas or liquid. Most can change from one to the other.

....

What change of state caused the bellows-type thermostat to open?

....

What change of state caused the wax type to operate?

....

Conduct an experiment to determine the serviceability of bellows-type and wax-type thermostats. Sketch below the equipment used and complete the table.

Thermostat make and type	Visual defects if any	Opening temperature °C	Closing temperature °C	Manufacturer's specified temp. °C	Serviceability

What is an advantage of the wax type compared with the bellows type?

....

....

Why is the opening of the bellows-type thermostat affected by cooling system pressure? ....

....

What is an advantage of the bellows-type thermostat compared with the wax type? ....

....

State the purpose of the small hole and pin in the valve disc.

....

....

COOLING SYSTEM PRESSURE

A cooling system is pressurised to .... the .... temperature of the coolant.

Why is such a design feature considered necessary?

....

What limits the pressure that can be imposed on a water-cooling system?

....

For approximately every 5 kPa of pressure rise the temperature increases 1°C.

What would the approximate boiling temperature of water in a cooling system be, using:

(i) a 4 lbf/in² (30 kPa) pressure cap? ....

(ii) a 15 lbf/in² (105 kPa) pressure cap? ....

189

ELECTRICAL SYSTEMS – TECHNOLOGY

THE D.C. COMMUTATOR-TYPE GENERATOR

Complete the drawing to show pictorially how the field poles and brushes are wound on a dynamo.

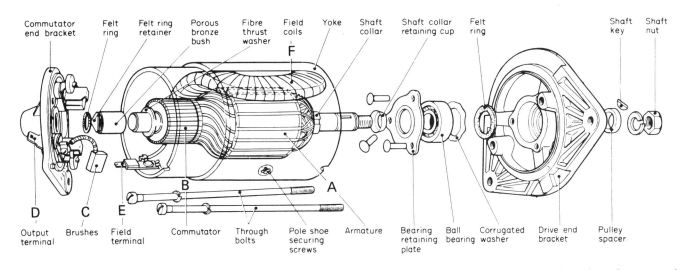

Commutator end bracket — Felt ring — Felt ring retainer — Porous bronze bush — Fibre thrust washer — Field coils — Yoke — Shaft collar — Shaft collar retaining cup — Felt ring — Shaft key — Shaft nut

F

D — C — E — B — A

Output terminal — Brushes — Field terminal — Commutator — Through bolts — Pole shoe securing screws — Armature — Bearing retaining plate — Ball bearing — Corrugated washer — Drive end bracket — Pulley spacer

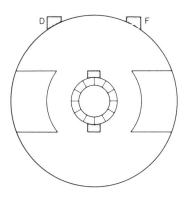

D   F

Explain the function of the lettered items shown in the diagram. The letters have been arranged so that the operation of the generator may be logically developed.

A ....

....

....

....

....

....

....

....

B ....

....

....

....

C ....

....

....

D ....

....

E ....

....

F ....

....

....

....

191

DYNAMO OUTPUT

The dynamo output must be sufficient to supply current that will maintain continuous electrical loads (when applied) and be capable of maintaining the battery in a satisfactory state of charge.

If the dynamo is suspected of being faulty a series of systematic checks should be carried out to determine its condition.

What two preliminary checks on the vehicle should be carried out before the dynamo itself is checked?

....

....

....

State what is actually being checked in the tests opposite?

Test (1) ....

....

Test (2) ....

....

DYNAMO FAULTS — Complete the fault table below:

Dynamo part	Possible fault
Armature	
Field coils	
Brushes	
Bushes	

INVESTIGATION

Check the electrical operation of a dynamo fitted either to a vehicle or an electrical test machine.

What two instruments are required to check the dynamo?

....

The two basic tests are listed below. Show the meter positions for each test and describe the test procedure.

1. TESTING THE ARMATURE CIRCUIT

....

....

....

....

....

....

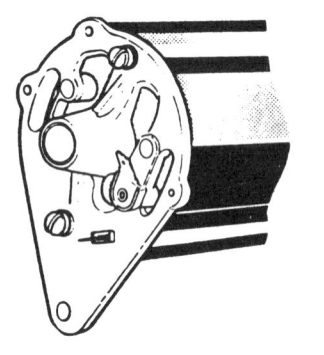

Results   Expected ....

Actual ....

2. TESTING THE FIELD CIRCUIT

....

....

....

....

....

....

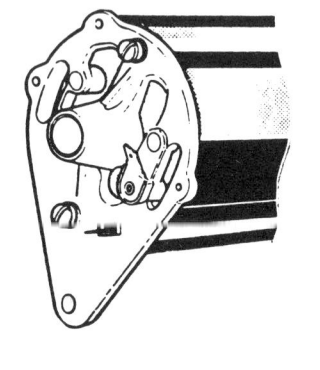

Results   Expected ....

Actual ....

METHODS OF GENERATOR CONTROL

The output of the generator must be controlled in order to prevent it from overloading itself. One method of control is the 'compensated voltage control' system.

Why would the dynamo overload itself if not controlled?

....

....

COMPENSATED VOLTAGE CONTROL

Voltage regulator

Name the main parts.

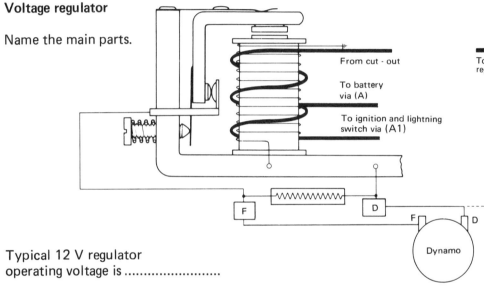

From cut - out

To battery via (A)

To ignition and lightning switch via (A1)

Dynamo

Typical 12 V regulator operating voltage is ....

With the aid of the diagram explain the function of the voltage regulator and how it may be adjusted.

....

....

....

....

....

....

....

Cut-out regulator

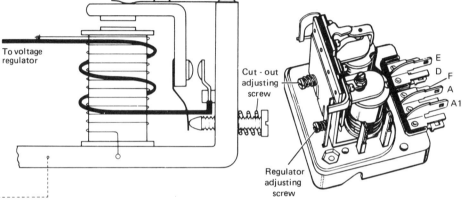

To voltage regulator

Cut - out adjusting screw

E
D
F
A
A1

Regulator adjusting screw

Typical 12 V system cutting in voltage.

....

Cutting-out voltage.

....

With the aid of the diagram explain the function of the cut-out regulator and how it may be adjusted.

....

....

....

....

....

CURRENT VOLTAGE REGULATORS

The current voltage system of control allows a discharged battery to be recharged initially at a uniformly high rate. The graph shows how the charge rate varies in the two systems.

Complete the diagram to show the wiring of a current voltage regulator.

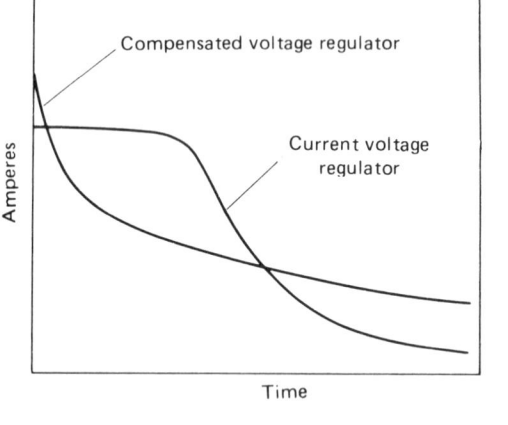

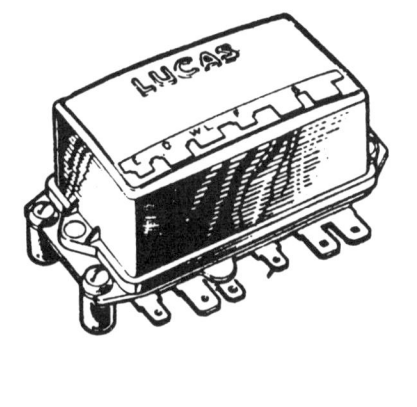

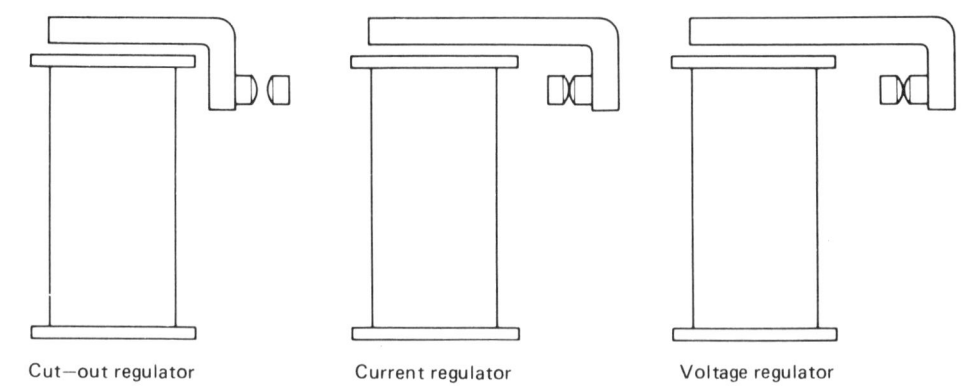

Cut—out regulator   Current regulator   Voltage regulator

Complete the circuit diagram to show the correct external wiring of a dynamo charging system.

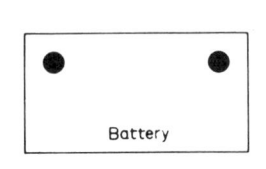

Battery

Solenoid switch

Control box (section and letter)

Ignition switch

Ignition warning light

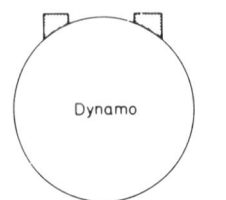

Dynamo

Explain with the aid of the diagram above, when completed, the operation of the current voltage regulator.

....
....
....
....
....
....
....
....
....
....

INVESTIGATION
TESTING A CURRENT VOLTAGE CONTROL UNIT

Before a regulator itself is checked, systematic checks should first be made on battery conditions, dynamo operation and continuity and connections of supply leads.

Show position of test meters in each sketch and describe the test procedure.

Note! Adjustment must only be carried out in conjunction with the infomation given on the meters.

Test 1. Voltage regulator setting.

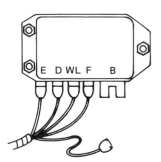

Test 2. Cutting-in voltage.

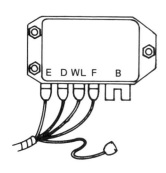

Reading obtained ....

....
....
....
....
....
Reading obtained ....

....
....
....
....
....
Reading obtained ....

Test 3. Current regulator setting. ....

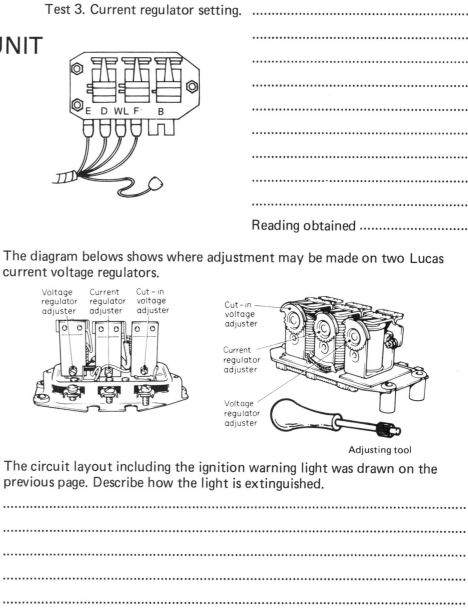

....
....
....
....
....
....
Reading obtained ....

The diagram belows shows where adjustment may be made on two Lucas current voltage regulators.

Voltage regulator adjuster   Current regulator adjuster   Cut–in voltage adjuster

Cut–in voltage adjuster
Current regulator adjuster
Voltage regulator adjuster

Adjusting tool

The circuit layout including the ignition warning light was drawn on the previous page. Describe how the light is extinguished.

....
....
....
....
....
....

THE ALTERNATOR

The alternator shown opposite consists of a series of magnets which are rotated in the centre of three sets of inter-wound coils.

Lucas
AC type

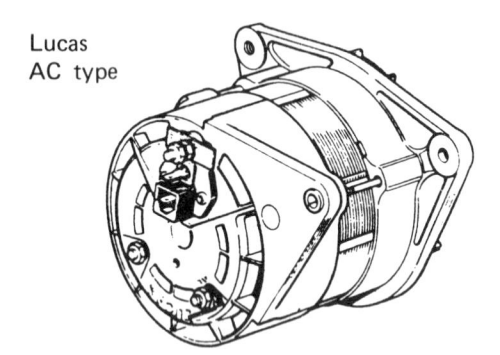

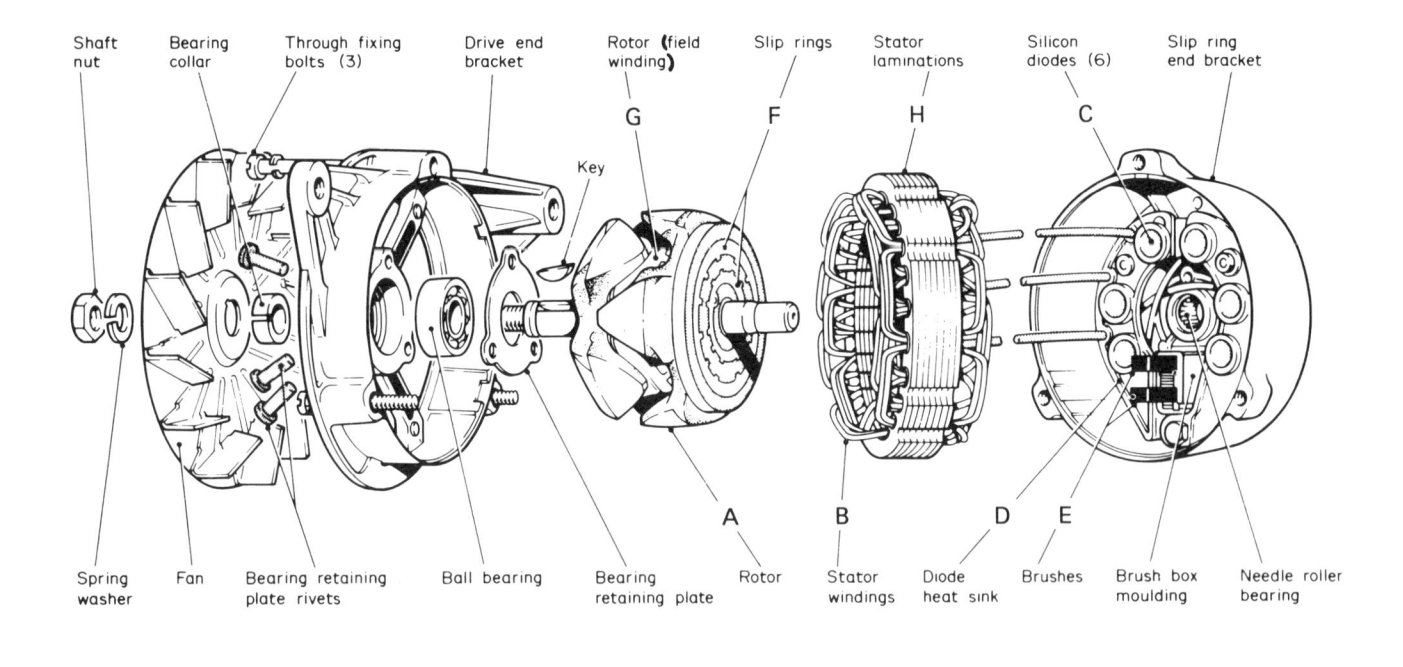

Explain the function of the lettered items shown in the diagram. The letters have been arranged so that the operation of the alternator may be logically developed.

A ....

....

....

B ....

....

C ....

....

D ....

....

E ....

....

....

F ....

....

....

G ....

....

....

H ....

....

Lucas ACR alternator

Name the major parts of the alternator shown below.

In what way does it differ from the one on the previous page?

....

....

....

....

....

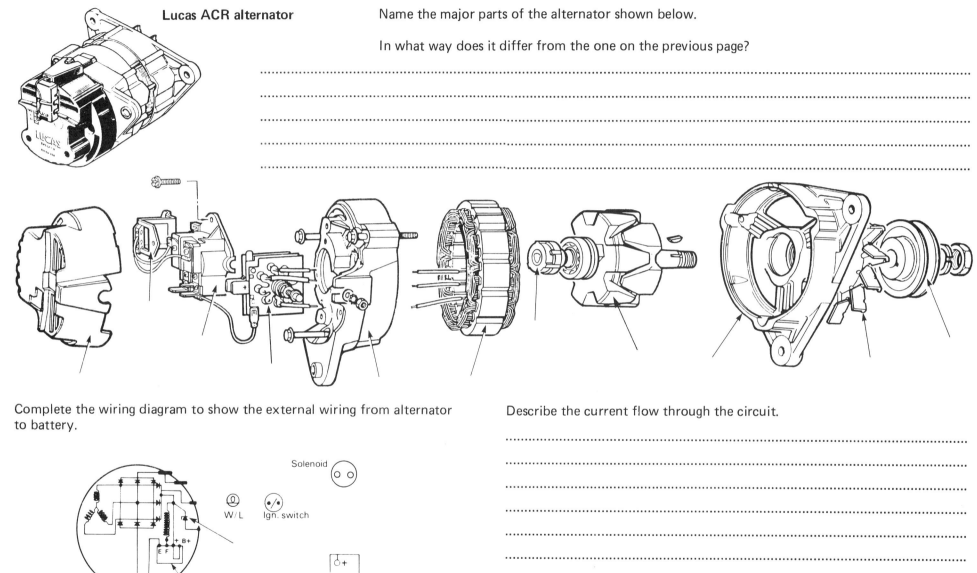

Complete the wiring diagram to show the external wiring from alternator to battery.

Solenoid

W/L   Ign. switch

Regulator

Describe the current flow through the circuit.

....

....

....

....

....

....

....

....

....

197

CHECKING ALTERNATOR SYSTEM

INVESTIGATION (Show position of meters in each case.)

List preliminary checks that should be made before alternator is tested.

....

....

Test 1. Alternator output

Expected output ....

Actual reading.................................

Test 2. Charging circuit volt drop

V_1 actual reading ....

V_2 actual reading ....

Test 3. Check alternator control.

Actual reading ....

ALTERNATOR FAULTS

Instant failure of rectifier (diode) pack can be caused by:

....

....

Instant failure of the regulator can be caused by:

....

....

State possible faults for the following conditions:

Low charge	
Warning light remains on	
No warning light	
High charge rates	
Alternator runs hot	

ELECTRONIC REGULATOR

The charging rate of alternators may be controlled by either a mechanical vibrating point voltage regulator, similar to the one described on p. 193, or by an electronic regulator.

In these units the point contacts have been replaced by a number of transistors and the size of the regulator is considerably reduced.

Explain with the aid of the simplified circuit diagram below how the e.m.f. voltage build up is controlled by the alternate operation of the two transistors aided by the avalanche diode.

....
....
....
....
....
....
....
....
....

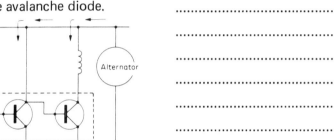

Below is shown the complete circuit which includes resistors to control current flow to the transistors, an extra transistor to reduce loading and a diode to protect the transistors from inductive voltage surges when T3 is switched off.

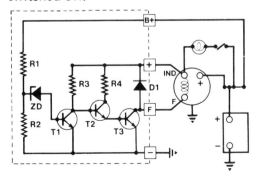

Electronic regulator-circuit diagram

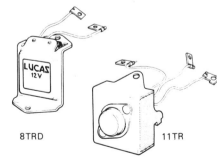

8TRD   11TR

Typical electronic regulators

COMPARISON OF DYNAMOS AND ALTERNATORS

Examine the dismantled components of the above machines and complete the table below. Give a general comparison of the items listed.

Item	Alternator	Dynamo
Make and model		
Output windings (position, action, size)	....	....
	....	....
Current rectified by	....	....
Field control windings (position, number of magnets)	....	....
Maximum rotational speed		
Current output		
Weight of machine		

Give the main advantages and disadvantages of using an alternator instead of a dynamo.

ADVANTAGES ....

....

....

DISADVANTAGES ....

....

199

STARTER MOTOR CONSTRUCTION

The starter motor shown opposite is of a conventional series parallel construction.

Trace the path of current flow through the motor by stating each component through which it passes.

Item	Current flow
1	*Feed terminal*
2	*Field coils*
3	
4	
5	
6	

Modern starters are mainly wound in some form of series pattern. That is, using a heavy strip field winding in series with the armature so that the field and armature current are equal.

....

....

....

....

....

....

Mark the diagram with the item numbers.

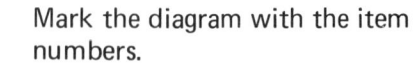

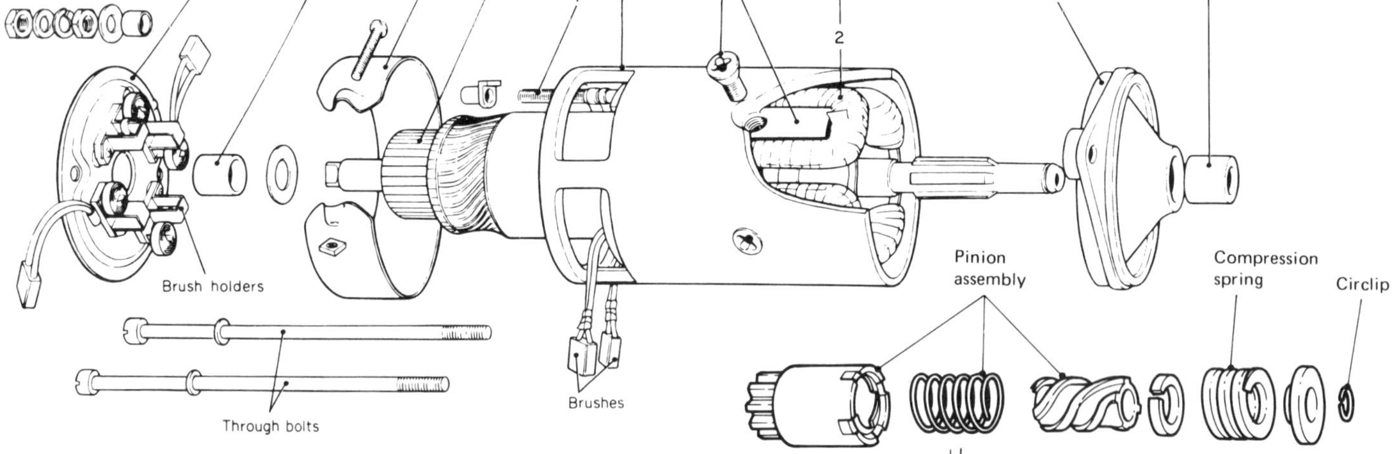

Terminal nuts and washers — Commutator end bracket — Bearing bush — Cover band — Commutator — Yoke — Pole shoes and screws — Field coils — Drive end bracket — Bearing bush — Brush holders — Through bolts — Brushes — Pinion assembly — Compression spring — Circlip

Show the wiring layout for the above starter motor and indicate the flow path through the circuit.

Series parallel

INERTIA DRIVE

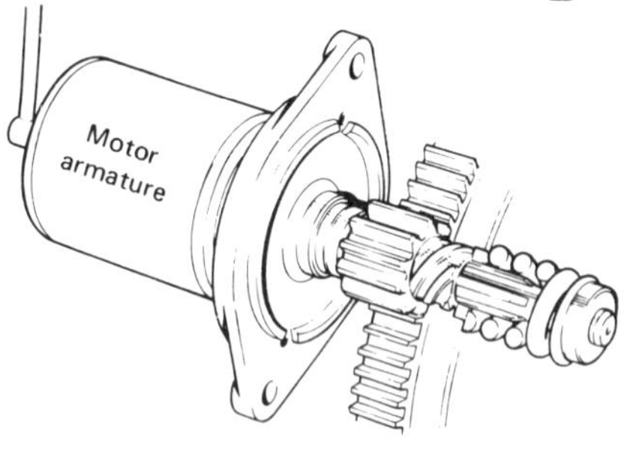

As shown, the drive consists of a pinion, mounted on the armature shaft.

Describe, with the aid of the drawings, its operation when the starter turns.

Motor armature

....

....

....

....

....

....

LUCAS M35J PRE-ENGAGED STARTER MOTOR

Name the
parts indicated.

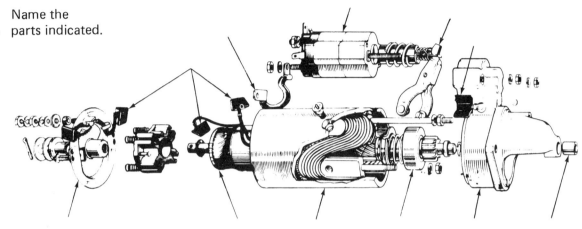

Complete the wiring diagram to show the electrical flow through the circuit.

The diagram shows a more modern design of starter motor than the one shown on the previous page.

State how the components named below differ

Field windings ....

....

....

Commutator ....

....

Brushes ....

....

Brush carrier ....

....

Describe the operation of the starter shown.

....

....

....

....

....

....

....

....

....

....

....

....

....

Starter motor

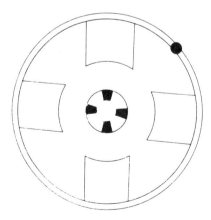

Solenoid

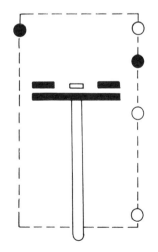

201

Name the indicated parts.

The type of starter motor shown left is

....

The drive on the previous page and the one on the left incorporate a one-way roller clutch assembly (freewheel).

What is the object of having this unit in the drive?

....

....

....

....

....

....

....

....

....

....

When the starter is required to transmit heavy torques a multi-plate clutch drive is often used (as shown below).

Describe how the parts are assembled together.

....

....

....

....

Explain how this type holds and disengages drive.

....

....

....

....

....

....

....

What are the advantages of the pre-engaged starter motor when compared with the inertia type?

....

....

....

....

....

....

....

Rivet | Pinion retaining ring | Barrel unit | Thrust washer | Backing ring | Clutch plates inner outer | Helical splines | Driving sleeve | Circlip | Lock ring

Pinion | Helical splined sleeve | Cushion spring | Ring nut | Pressure plates | Shim | Moving member | Retaining washer | Engagement bush

TESTING STARTER MOTORS

If the starter motor is considered suspect, i.e. it is cranking the engine, but only slowly, a systematic check should be carried out to determine if there is an excess voltage loss (high resistance) in the circuit.

INVESTIGATION

Carry out a series of voltage checks to determine the condition of a starter circuit using a 0–20 V voltmeter.

Before tests are carried out preliminary checks that must be made are:

....

....

....

In all six tests shown the engine must be cranked without starting. This is achieved by:

S.I. engine ....

C.I. engine ....

Show position of voltmeter for each test.
State expected and actual readings.

Test	1	2	3	4	5	6
Expected voltage						
Actual voltage						

1. Battery voltage on load

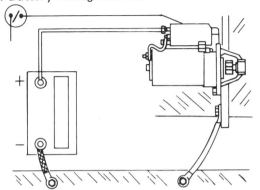

2. Voltage at solenoid operating terminal.

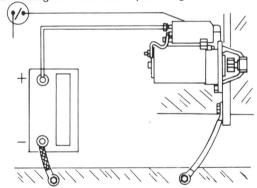

3. Voltage at starter on load.

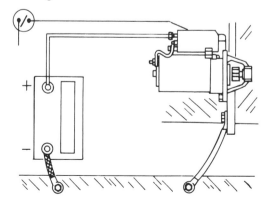

4. Voltage drop insulated link

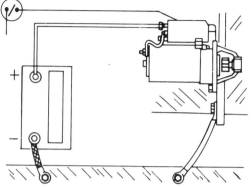

5. Voltage drop solenoid contacts.

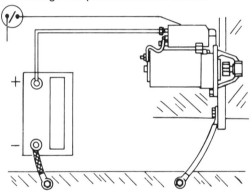

6. Voltage drop earth line.

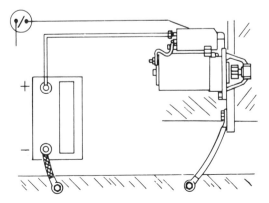

AXIAL STARTERS

The starting of heavy vehicle engines both petrol and (especially) diesel poses special problems, mainly due to the much larger compression loads. The diesel engine also requires to be motored for longer than the petrol engine in cold temperatures.

The axial starter shown (semi-pictorially) below employs additional auxiliary winding and a holing coil as well as the main coil.

....

....

INVESTIGATION

Examine a starter of this type, observe its operation and answer the questions opposite.

In the position illustrated, the starter switch has just been closed. Show the field poles, and name the indicated parts.

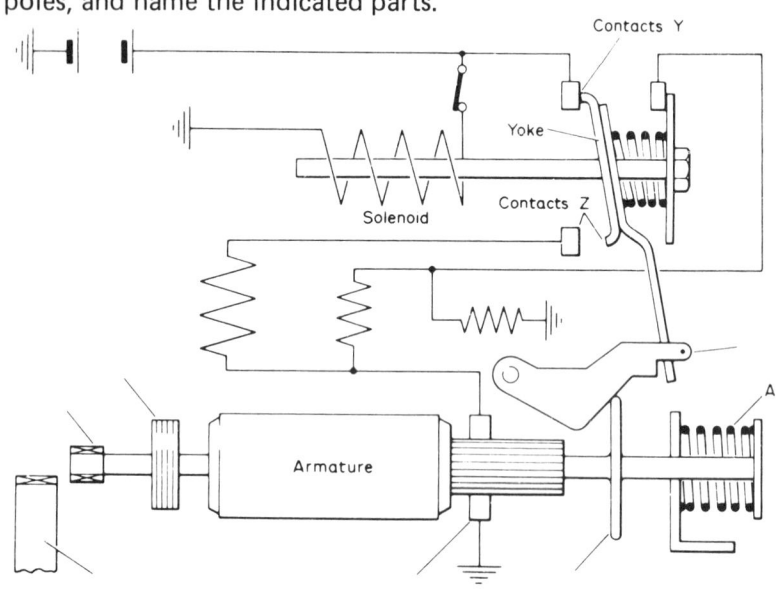

1. In what position are contacts Y and Z when the starter switch is open?

....

2. Explain briefly what happens initially when the starter switch is closed.

....

3. Which windings are energised in this condition (shown opposite)?

....

....

4. What effect does this have on the armature? ....

....

5. What happens to the contacts as a result of this? ....

....

6. How does this affect the main and auxiliary windings?

....

....

7. What is the purpose of the holding coil?

....

....

8. What is the purpose of the spring A shown on the right-hand end of the armature?

....

....

9. Why is a multi-plate clutch used between the pinion and armature shaft?

....

....

COAXIAL STARTERS

The coaxial starter differs from the axial starter in that only the pinion moves axially.

....
....
....
....
....

From battery

From starter switch

At rest position

INVESTIGATION

Examine a coaxial starter and answer the following questions.

1. What provison is made to cause the armature to rotate slowly at reduced voltage during initial engagement of the pinion?

....
....

2. Why is a helix required on the pinion shaft? ....

....

3. How is full voltage applied to the armature when the pinion is fully engaged?

....
....

4. What is the function of the ball locking device?

....
....

The drawings show the pinion engagement sequence of a CAV coaxial starter. Label the parts as required and explain alongside each drawing what is happening.

....
....
....
....
....

A

....
....
....
....
....
....

B

....
....
....
....
....
....
....

C

....
....
....
....
....

ADVANTAGES OF HEAVY DUTY STARTERS

An advantage of the axial starter motor is:

....

....

An advantage of the coaxial starter motor is:

....

GENERAL STARTER MOTOR FAULTS

Give faults that can cause the following starter motor symptoms.

Symptom	Faults
Starter fails to turn engine	....
	....
	....
	....
	....
	....
	....
Starter turns engine very slowly	....
	....
	....
Starter noisy or excessively rough in engagement	....
	....
	....
	....
	....

LIGHTING SYSTEMS

Vehicle lights may be divided into two classes, those that are obligatory and those which may be considered as accessories, although they may be standard fittings on the more expensive type of vehicle.

HEADLAMP SYSTEMS

Vehicles use either sealed-beam-type units or pre-focus, semi-sealed units. The bulbs are mainly of a twin-filament design and positioned so that they emit either a horizontal (main) beam or a dipped beam; modern systems usually employ halogen bulbs.

TWIN-PAIR HEADLAMPS

Headlamps when on main beam are expected to provide (1) a wide beam, illuminating the area directly in front of the vehicle and (2) a long, searching beam capable of picking out objects in the distance. It is because of these two opposing requirements that twin-pair headlamps have been developed.

Which lamps operate when on dip beam?

Complete the wiring diagram to show a twin-pair headlamp arrangement; add filaments to the headlamps as necessary and name parts.

State the colour of the cables.

HEADLAMP ALIGNMENT

All vehicle headlamps in the UK must comply with the Ministry of Transport Road Vehicle's Headlamp Regulations, which state the position, or angle, of the dipped beams.

The specialist equipment used to align headlamps measures the angle of dipped beam and the beams' positions relative to one another when both on main and dipped beam.

Show a sketch of such equipment in its testing position.

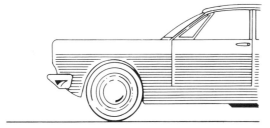

Type of gauge

What pre-checks are necessary to ensure that headlamp alignment is accurately carried out?

INVESTIGATION

Check the headlamp alignment of a vehicle, using available equipment, and describe the alignment procedure.

Vehicle Make .... Model ....

....

....

....

....

On the aiming screens shown indicate correct main and dipped beam positions.

1. Typical main driving beam position all types.

....

....

....

....

....

....

....

....

Vertical screen | Aiming line

Horizontal screen
Aiming line

2. Symmetric dipped (passing) beam.

....

....

....

....

....

....

3. Asymmetric dipped (passing) beam.

....

....

....

....

....

....

DIRECTION INDICATORS

There are various methods of breaking the electrical circuit to the lamps and so causing them to flash. The two most common being the hot-wire and the vane-type flasher units.

....

....

HOT-WIRE FLASHER UNIT

This is the original type of flasher canister. It is rarely used on modern cars. Complete the wiring circuit and describe its basic operation.

O
P
OL BO

P   B   L

....
....
....
....
....
....

VANE-TYPE UNIT

This unit is much simpler in operation than the other type.

The base supports a snap-action metal vane held in tension by a metal ribbon.

When not in operation the contacts are closed.

Name the main parts, and indicate current flow.

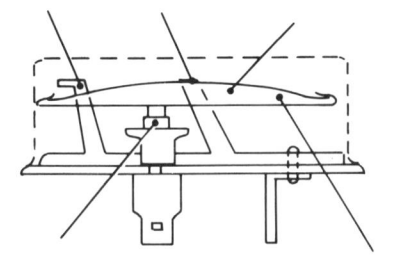

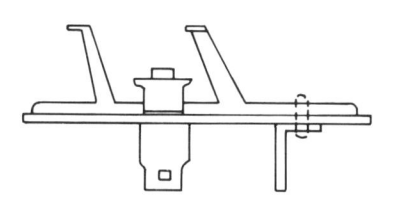

Show position of ribbon and vane when contacts are open.

Describe its operation.

....
....
....
....

What usually occurs to the other lamps if one bulb fails on the
(i) hot-wire unit?

....

(ii) vane unit? ....

(iii) hazard warning unit?

....

The legally required flashing rate is between ....

208

Sketch simple wiring diagrams to represent the following circuits:

DIRECTION INDICATOR

Include flasher unit panel indicator lamps and, if required, side indicators.

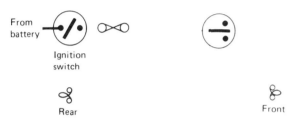

From battery

Ignition switch

Rear

Front

SIDE AND TAIL LIGHT CIRCUIT

Side marker lights should be shown if a heavy commercial vehicle is used as an example.

Rear

Front

All lights are wired in ....

STOP LAMPS

Indicate how they are operated.

Ignition switch

REVERSING LAMPS

These lamps must operate only when the vehicle is being driven in reverse. Show how the switching arrangement may be designed to ensure this always occurs.

Ignition switch

The rear lights, stop lights and flashers of many vehicles are connected via a resistor.

What is the purpose of this unit?

....

....

....

HAZARD WARNING FLASHER UNIT

This unit when operated flashes all indicator lamps simultaneously to provide external indication of a potential traffic hazard.

The contacts are open when not in operation. Its operational sequence is similar to the hot-wire type described on the previous page.

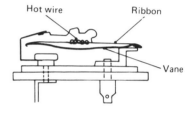

Hot wire   Ribbon

Vane

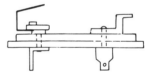

Show position of ribbon and vane when contacts are closed.

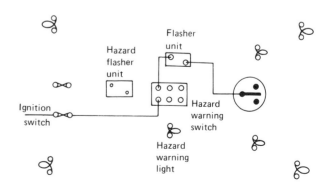

Flasher unit

Hazard flasher unit

Ignition switch

Hazard warning switch

Hazard warning light

LIGHTING FAULTS

A vehicle's lights, when used on the road, must be maintained in good working order at all times, day or night. Faults, therefore, must be immediately rectified.

Lighting defects fall into two groups.

State possible symptoms and causes.

1. Optical condition.

....
....
....
....

2. Mechanical/electrical operating condition.

....
....
....
....
....

The less obvious symptoms require the use of a systematic volt-drop check.

....
....
....

VOLT DROP CHECK

When checking for high-resistance (bad) connections a volt-drop check should be made.

State what is being checked on the diagrams and carry out a similar check.

Give expected and actual values.

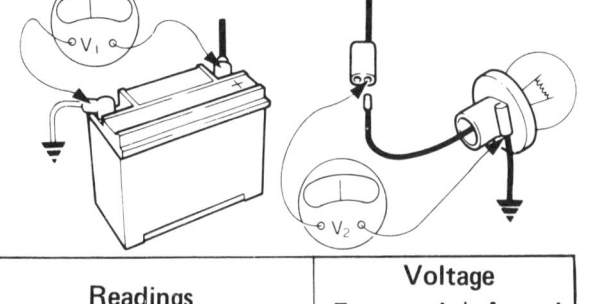

| Readings | Voltage | |
	Expected	Actual
V₁		
V₂		
V₃		
V₄		
V₅		

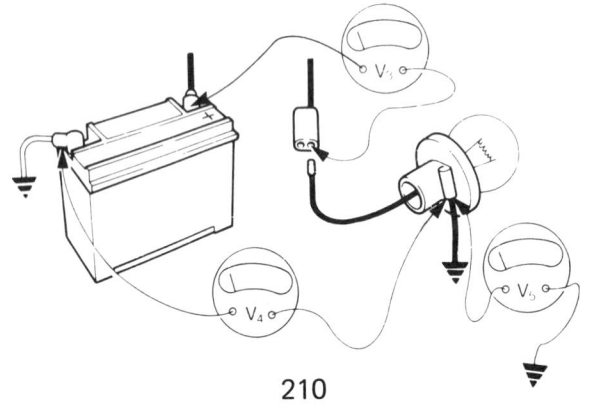

CHECKING INSULATED LINE VOLT DROP

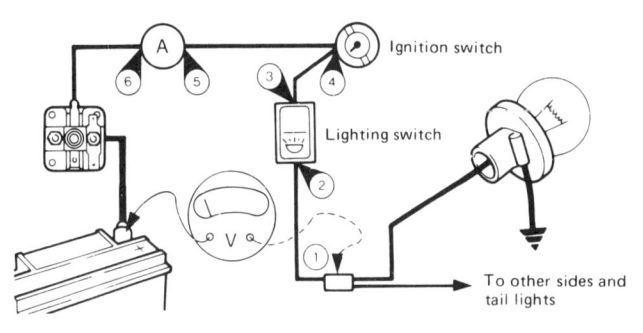

Describe the basic procedure.

....
....
....
....
....

How should a check for open circuits be made?

....
....
....
....
....
....
....

TRAILER/CARAVAN ELECTRICS

All types of trailer units (including caravans) must show the obligatory rear-facing lamps. These must be operated from the driving vehicle's battery and not a separate battery (i.e. that could be fitted to the trailer to provide interior lighting).

In order to provide this supply safely, and allow it to be easily disconnected from the vehicle, 7-pin connectors are used. An example is shown below.

Identify each part.

 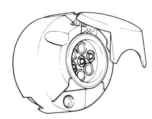

On older vehicles and caravans a single 7-pin connector is able to supply all the electrical needs. On modern caravans it is obligatory to fit rear fog lights. This extra item, together with the possible fitting of reversing lights and internal caravan electrical equipment, has made the fitting of two 7-pin connectors essential.

The first (or existing) connector is known as a ....

The second connector is known as a ....

Number each connection on the drawings and indicate how the sockets are arranged to prevent interchangeability.

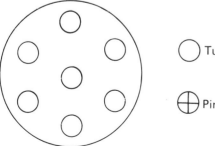 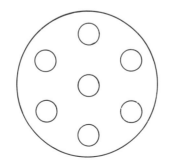

The vehicle's rear cable harness arrangement shows the items that should be connected to the 12 N socket. With the aid of the table below, complete the diagram to show the socket correctly wired.

Number the socket connections and indicate the cable colours.

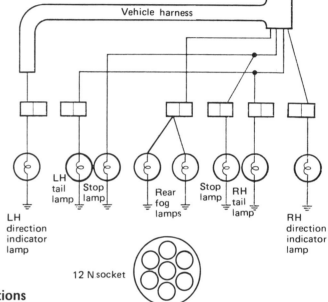

Pin dispositions

Pin no.	7-core cable colour	Circuit allocation	
		12 N connector (ISO 1724)	12 S connector (ISO 3732)
1	Yellow	LH flashers	Reverse and/or mechanisms
2	Blue	Rear fog (auxiliary, older vehicles)	No allocation (additional power)
3	White	Common return (earth)	Common return (earth)
4	Green	RH flashers	Power supply (caravan interior)
5	Brown	RH side/tail/no. plate	Sensing device (warning light)
6	Red	Stop	Power supply (refrigerator)
7	Black	LH side/tail/no. plate	No allocation

HEAVY-VEHICLE 7-PIN CONNECTIONS

As with caravan electrics, a single 7-pin plug and socket assembly is insufficient to meet the present and future tractor/trailer requirements for 24 V commercial vehicles.

Two 7-pin plugs are recommended for use — these are known as ....

.........................., i.e. the original 7-pin arrangement, and........................

.... to cater for additional circuits.

Number each connection on the drawings and indicate how these sockets are arranged to prevent interchangeability.

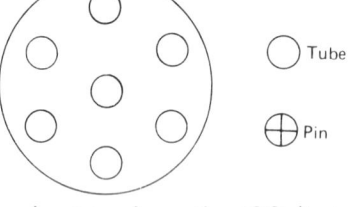
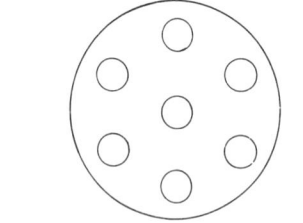

Complete the chart to show the ISO (International Organisation for Standardisation) recommended circuit allocations.

Pin no.	7-core cable colour	Circuit allocation	
		24 N connector	24 S connector
		(ISO 1185)	(ISO 3731)
1	White		
2	Black		
3	Yellow		
4	Red		
5	Green		
6	Brown		
7	Blue		

When connecting articulated trailers, helical cable similar to the one shown below is used. It is available in 5-, 7- or 10-core. The cables are PVC-insulated and protected in a strong nylon sheath.

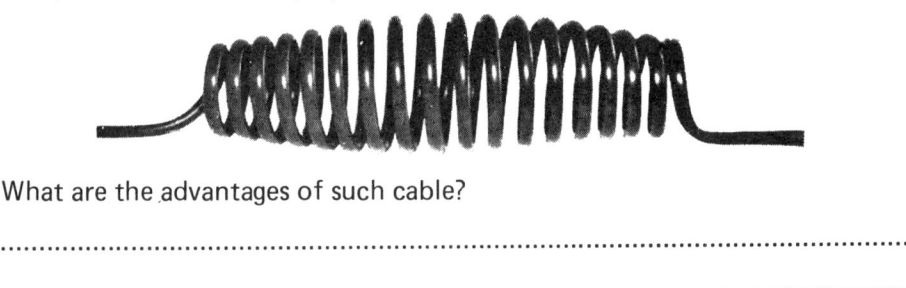

What are the advantages of such cable?

....

....

....

INSULATED RETURN WIRING SYSTEM

Whilst most cars and light commercial vehicles employ earth-return electrical systems, many heavy vehicles and all public service vehicles use insulated return systems.

What are the main advantages of this system?

....

....

....

INTERIOR LIGHTING FOR PSVs

Conventional bulbs are frequently used for the interior lighting of public service vehicles, but fluorescent tubes are rapidly becoming more popular. One reason for this is the development of transistors and thyristor circuits which convert d.c. current to a.c. current to operate these tubes.

What are the main advantages of fluorescent lamps?

....

....

WINDSCREEN WIPERS

The wipers are operated by a mechanism which is given a to-and-fro action by a suitably designed linkage which is connected to the armature shaft of a small electric motor.

Identify the types
of drive assembly.

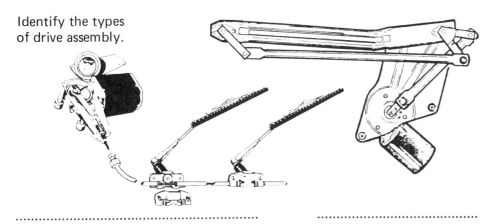

....

The electric motor may be of a single-twin or variable-speed type. Almost all types have a limit-switch device incorporated into the drive assembly.

Below is shown the wiring diagram of a single-speed windscreen wiper motor. Explain the operation of the limit-switch when the main control switch is opened.

....

....

....

....

....

....

....

....

Windscreen
wiper motor

Control
switch

Fuse

Limit
switch

Ignition
switch

From
battery

Describe the drive arrangement on both motors and state if single or twin speed, giving reason for choice.

....

....

....

....

....

....

Wiper is .... speed; there are .... brushes.

....

....

....

....

....

Wiper is .... speed; there are .... brushes.

213

HORNS

With the aid of this simple diagram explain the operation of the horn.

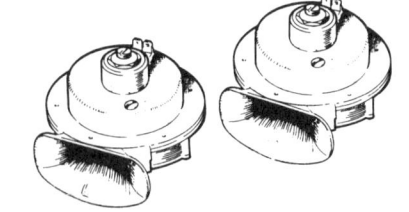

Name the various parts.

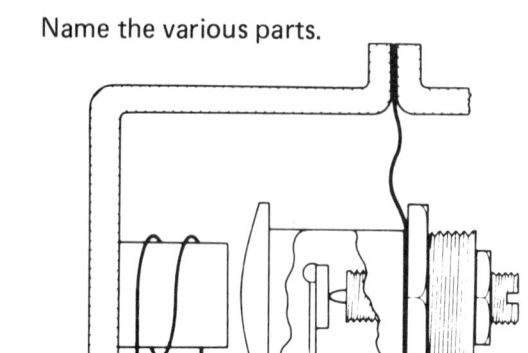

Terminals

....
....
....
....
....
....
....
....
....
....

Complete the wiring diagram of the simple twin-horn circuit shown below.

From
battery

Horn
push

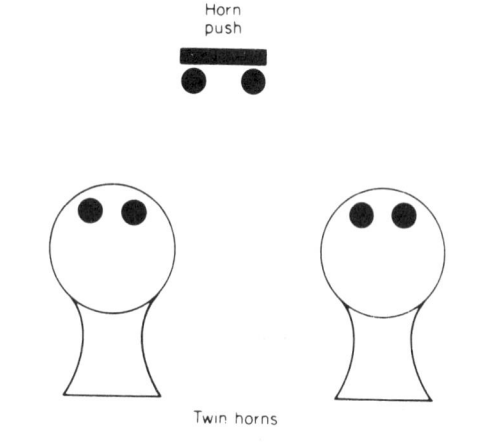

Twin horns

RELAY SWITCHES

When electrical components — such as horns — use a large amount of current a relay may be used.

....
....
....

Complete the wiring of the three-terminal relay shown and explain its principle of operation.

Indicate terminal connections.

....
....
....
....
....
....
....
....
....

Draw.a wiring diagram of a twin-horn circuit to include a relay switch.

214

BI-METAL FUEL AND TEMPERATURE GAUGES

These gauges are fitted on most modern vehicles. The current supply to both fuel and temperature gauge is controlled by a voltage stabiliser. The sender units in both fuel tank and cooling system have resistors that vary depending upon either the amount of fuel in the tank or the engine temperature.

Explain the principle of operation of the:

voltage stabiliser

....

....

....

....

....

bi-metal gauge

....

....

....

....

....

engine cooling system temperature transmitter

....

....

....

....

See p. 90 for operation of oil warning light.

Complete this wiring diagram to show the internal wiring symbols of the voltage stabiliser, gauge units and temperature and fuel gauge transmitter units.

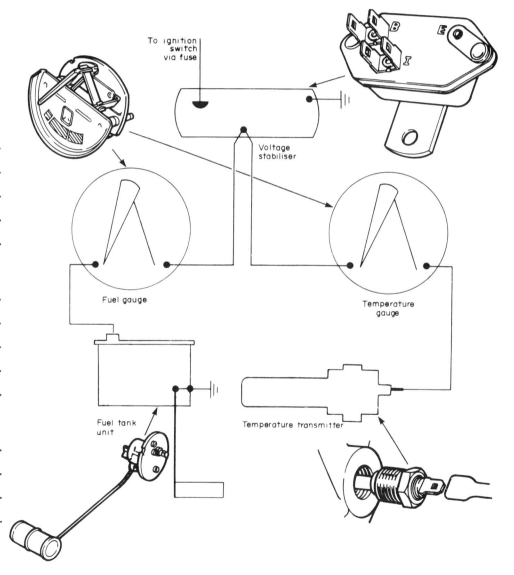

To ignition switch via fuse

Voltage stabiliser

Fuel gauge

Temperature gauge

Fuel tank unit

Temperature transmitter

215

CAR AND CAB HEATER MOTORS

Heater motors are normally designed to operate at either two speeds or through a variable speed range. Complete the diagrams to show:

1. Heater motor having variable speed.

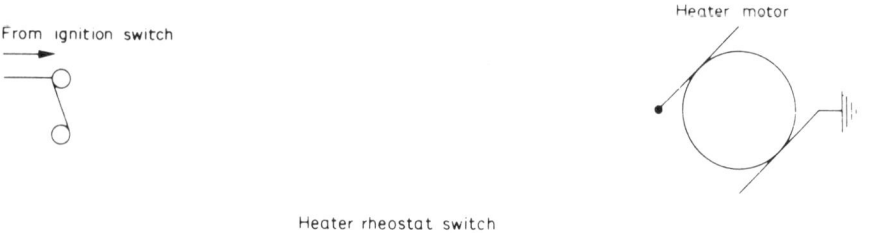

From ignition switch

Heater motor

Heater rheostat switch

2. Heater motor controlled through two-speed switch.

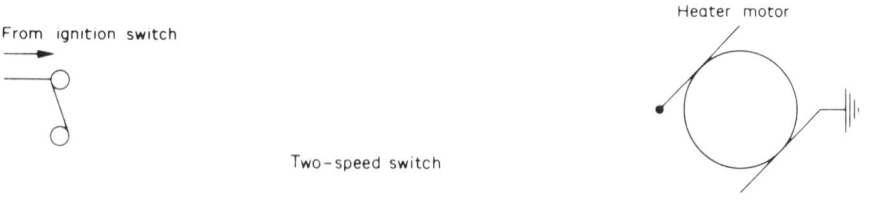

From ignition switch

Heater motor

Two-speed switch

INSTRUMENT PANEL

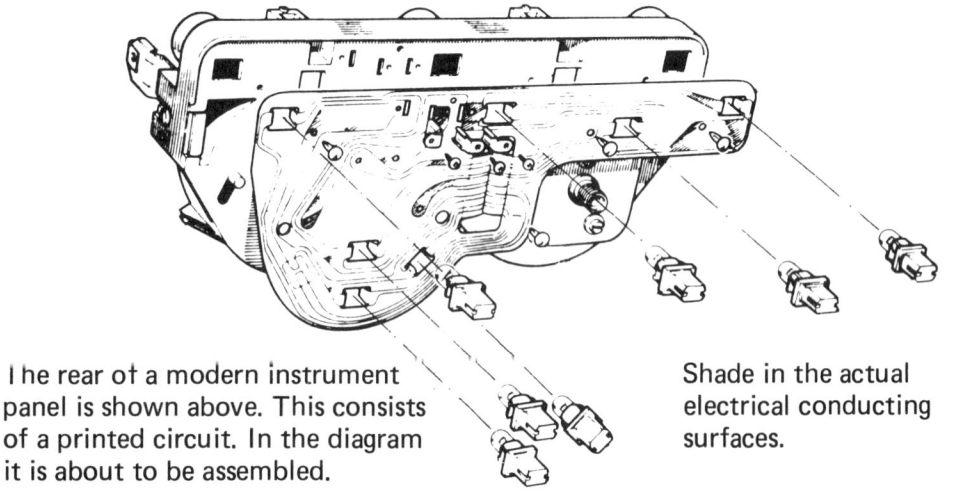

The rear of a modern instrument panel is shown above. This consists of a printed circuit. In the diagram it is about to be assembled.

Shade in the actual electrical conducting surfaces.

CABLE LAYOUTS

The diagram shows the under-dash wiring harness of a light commercial vehicle

Name the items indicated.

CABLE SIZE

For practical purposes the electrical resistance of a cable depends on:

(a) .... (b) .... (c) ....

What may occur if cables of insufficient size were fitted to a vehicle?

....

In general what is the maximum permissible voltage drop in circuits?

....

Complete the table to indicate typical cable size and capacity.

Circuit	Cable size	Cross-sectional area approx.	Max. current capacity
Ignition L.T. accessories			
Headlamps			
Charging			
Starting			

LEAD–ACID BATTERY – CONSTRUCTION

Most vehicles use a 12 V lead–acid battery.

The battery shown below has a translucent polyproplene case and an 'Aqualok' one-piece manifold topping-up feature.

Note also the internal cell connections.

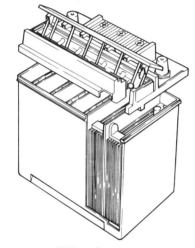

Automatic filling feature.

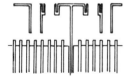 

Show position of electrolyte and stoppers:

1. after topping up vent cover open as on battery above.

2. when vent cover is replaced.

The components which make up one cell are shown below.

Name the parts and describe their construction.

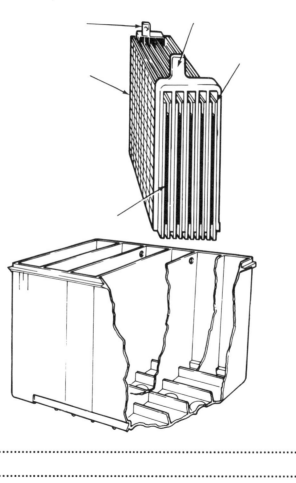

....

....

....

....

....

....

....

....

....

....

....

....

....

....

....

....

....

....

....

....

....

What active material forms the:

positive plate? ....

negative plate? ....

Why is there one less positive plate than negative plate?

....

....

....

....

What important maintenance feature do some more modern battery designs incorporate?

....

....

....

COMPOSITION OF PLATES AND ELECTROLYTE

A lead—acid cell when in a charged condition contains two dissimilar electrical conductors immersed in an electrolyte. In this condition it is able by chemical action to deliver an electric current until it becomes totally discharged.

Complete the charge/discharge sequence shown below.

Fully charged

| Positive plates | | Electrolyte | | Negative plates |

Lead dioxide
P_b O_2

Composition of plates and electrolyte

P_b SO_4
Lead sulphate

| Positive plates | | Electrolyte | | Negative plates |

Discharged

....
....
....
....
....

Why should batteries never be completely discharged?

....
....
....
....
....

Show by means of a graph the effect of the charging cycle on specific gravity and cell voltage.

Electrolyte Sp. Gr. values

1.290
1.230
1.170

1.8   2.0   2.2

Cell voltage (e.m.f.)

....
....
....
....
....
....

During the charge process the chemical reaction causes the plates to give off bubbles. Name the gas given off at the:

positive plates .... negative plates ....

State effects caused by excessive overcharging.

....
....
....

State effects caused by lack of charging or battery standing idle.

....
....

BATTERY FAULTS

State possible faults and causes which could create the following battery symptoms.

Battery symptom	Fault	Possible cause
Will not crank engine first thing in the morning but is O.K. after starting		
Always requires topping up		
Bubbles excessively and electrolyte is a dirty brown colour		
Car just stopped, all electrical power went off		
Will not accept charge		

BATTERY MAINTENANCE

List items of maintenance that should be carried out to prolong battery life.

....

....

....

....

METHODS OF CHECKING BATTERY CONDITION

If a battery is suspected of being faulty, a systematic check of its condition should be carried out. The battery must be in a reasonable state of charge in order that it can supply a heavy starting current for a short time.

Investigation

Carry out or state the basic tests or readings that should be made under the following headings.

VISUAL ....

....

HYDROMETER TEST

Indicate specific gravity reading on float shown.

....

Readings obtained

Cell number	1	2	3	4	5	6
Spgr value						

Discharged

70% charged

Charged

HIGH-RATE DISCHARGE TEST
This is made to ensure that each cell will supply the heavy currents required for starting.

....

Actual reading .... Expected reading ....

Any other effects ....

THREE-MINUTE BATTERY TEST ....

....

Actual reading .... Expected reading ....

THE ALKALINE BATTERY

The nickel—cadmium cell battery differs in many ways from the lead—acid type.

Complete the table below to describe the basic battery construction.

Component	Material
Cell casing	
Separators	
Plate	
Positive active material	
Negative active material	
Electrolyte	

What is the (nominal) voltage of each cell? ....

How many cells make up the following batteries?

(a) 6 V .... (b) 12 V .... (c) 24 V ....

What is the specific gravity range of the electrolyte? ....

To what extent does the specific gravity vary with the state of charge?

....
....

Maintenance — briefly state the requirements.

....
....
....

What is the effect of ageing on the electrolyte?

....
....
....
....
....
....

How should the effects of alkaline electrolyte be neutralised if spilt?

....
....
....
....
....

What are the relative merits of the alkaline battery compared with the lead—acid type?

....
....
....
....

What restricts the use of this type of battery?

....
....
....

ELECTRICAL SYSTEMS – APPLIED STUDIES

L18:36,37
L18:39,40

OHM'S LAW

'Ohm's Law', is the expression that relates voltage, current and resistance to one another.

State what these electrical terms mean, and state the units in which they are measured.

Voltage ....
....

Current ....
....

Resistance..
....

One of the relationships defining Ohm's Law states that providing the resistance is kept constant the current will double if the voltage is doubled. Expressed more mathematically this could be stated as:

....

....

Expressed as a formula using electric symbols

$$I = \frac{E}{R}$$

Where I = ...
R = ...
E = ...

(*Note.* The symbol V may be used instead of E.)

The formula may be   E = ... and   R = ...
transposed to state

EXAMPLE

Calculate the current flowing in a circuit when a pressure of 12 V is applied across a 3-ohm resistance.

Simple Ohm's Law problems

1. Calculate the current flowing in a coil of 4-ohm resistance when the electrical pressure is 12 V.

2. Calculate the voltage required to force a current of 2.5 amps through a resistance of 5 ohms.

3. A dynamo produces a current of 35 A when the voltage is 14.

   What is the resistance of the dynamo?

4. What voltage will be required to cause a current flow of 3 A through a bulb having a filament resistance of 4.2 ohms?

5. What will be the total resistance offered by a lighting circuit if a current of 11 A flows under a pressure of 13 V?

6. Two 12 V headlamp bulbs each have a resistance of 2.4 ohms.

   Calculate the current flowing in each bulb and the total current flowing in the circuit.

SERIES CIRCUITS

With the aid of diagrams state the basic laws of series circuits.

Voltage ....
....
....

Current ....
....
....

Resistance ....
....
....

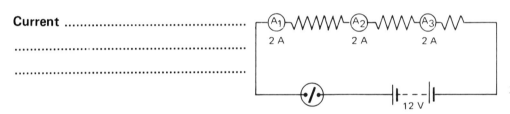

Problems

1. Two resistances of 1.75 Ω and 4.25 Ω are connected in series.

   What voltage would be required to cause a current of 2.5 A to flow in the circuit?

2. Four resistors of 6, 8 10 and 12 Ω are connected in series to a 12 V circuit.

   Calculate the total resistance of the circuit and the current flowing in each resistance.

3. Four resistors of equal value are placed in series and connected to a 110 V supply. A current of 5 A then flows.

   Calculate the value of each resistor and the voltage across each resistor.

4. Three resistors are wired in series and when connected to a 12 V supply a current of 6 A flows in the circuit.

   If two of the resistors have values of 0.5 and 0.8 Ω calculate the value of the third resistor.

5. Three resistors of 2, 4 and 6 Ω are connected in series to a 12 V battery.

   Calculate the current flowing in the circuit and the voltage across each resistance.

PARALLEL CIRCUITS

With the aid of diagrams state the basic laws of parallel circuits.

Voltage ....
....
....
....

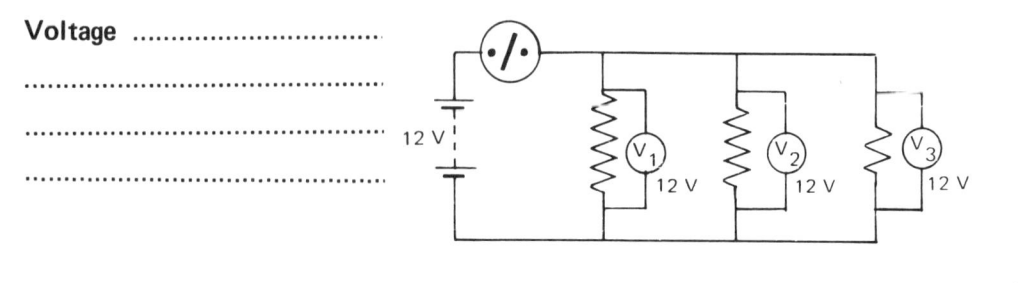

Current ....
....
....
....
....
....
....

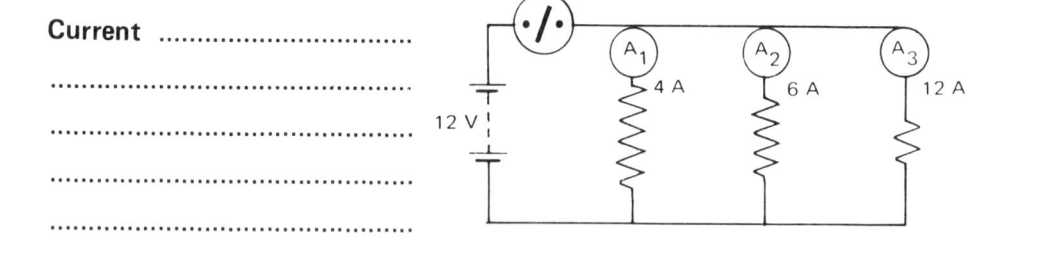

Resistance ....
....
....
....
....
....
....
....
....
....
....

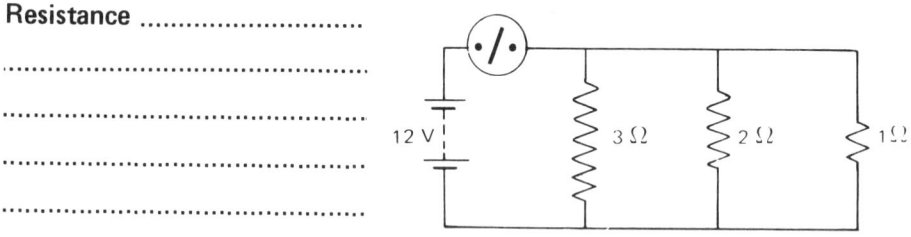

Problems

1. Three conductors are placed in a parallel circuit, their resistances being 2, 3 and 4 Ω.

   What current will flow in each when connected to a 12 V system?

2. Two resistances of 20 and 5 Ω are connected in a 12 V parallel circuit.

   Calculate the total current flow.

3. Two resistances of 8 and 6 Ω are connected in parallel.

   What voltage would be required to cause a current flow of 7 A?

4. Four resistors of 6, 8 10 and 12 Ω are connected in parallel to a 12 V circuit.

   Calculate the current flowing and the total resistance of the circuit.

5. Three resistors of 3, 5 and 6 Ω are connected to a 12 V battery.

   Calculate the total circuit resistance.

ELECTRICAL POWER

Power is the rate of doing work and is measured in ....

The definition of electrical power is: The definition of resistance is:

INVESTIGATION

Determination of lamp bulbs current and resistance.

Note. The wattage of each bulb must be different.

1. Calculate the current output of four bulbs and then their filament resistance.

Type of bulb	Wattage	Voltage	Current	Resistance

Calculations

Problems

1. Calculate the amount of current used by a 12 V 48 W headlamp bulb.

2. Calculate the amount of current used by two 12 V 54 W headlamp bulbs.

3. Calculate the amount of current used by five 12 V 5 W side lamp bulbs.

4. Calculate the total current flow when two 12 V 60 W headlamp bulbs, five 12 V 6 W side lamp bulbs, two 12 V 21 W flasher bulbs and three 12 V 3 W panel-light bulbs are in operation at the same time.

5. A motor vehicle operating on a 12 V system has the following lamps operating:

   Two 54 W headlamps.
   Five 6 W side and tail lamps.
   Two 48 W spotlamps.

   If the alternator is generating a current of 35 A, determine the ammeter reading, assuming the ignition system takes 6 A.

VOLTAGE DROP

In all electrical systems the voltage (or pressure) across the component being operated — say to starter motor — is always less then the voltage at the generating source, in this case the battery. This is because some electrical pressure is required to force the current along the cable from the battery to the starter motor.

Define:

e.m.f. ....

....

p.d. ....

....

v.d ....

....

The voltage drop in a circuit   v.d. =

Example

The open circuit battery voltage is 12 V. When the starter motor is operated the battery voltage is reduced to 9.5 V; determine the v.d.

v.d. =

Alternatively if the internal resistance of a component is known, the v.d. may be found by applying Ohm's Law.

$$V \quad = \quad I \quad \times \quad R$$

or v.d. =

Example

Calculate the v.d. in a starter cable when the internal resistance of the cable is 0.0012 Ω when carrying a current of 220 A.

V =

Problems

1. A battery has a p.d. of 9.75 V when operating a starter motor. The battery e.m.f. is 12.5 V.

   Determine the v.d. when the starter is in operation.

2. If the volt drop in a starter cable is 2.75 V and the e.m.f. 12.85 V, determine the terminal voltage at the starter motor when in operation.

3. Calculate the v.d. in a starter cable when the internal resistance of the cable is 0.0016 Ω when carrying a current of 284 A.

4. A 12 V battery gave an open-circuit voltage of 12.77 V.

   When supplying a current of 18 A the terminal voltage drops to 12·05 V.

   Calculate the internal resistance of the battery.

5. The leads from a 12 V battery to a starter motor have a total resistance of 0.005 Ω, what p.d. will be required to send a current of 200 A through these leads? What would be the voltage at the starter motor terminals in this case?

RESISTANCE OF ELECTRICAL CABLES

If the length of cable supplying current to a component is doubled the internal resistance of the cable will...............................

i.e. length is directly ....
.... to the resistance.

If the cable's cross-sectional area is doubled the internal resistance will ....

i.e. cross-sectional area is ....
.... to the resistance.

Problems

1. If the resistance of a wire is 0·009 Ω when it is 6 m long, what will be its resistance when it is 72 m long?

2. If 5 m of starter cable has a resistance of 0·025 Ω calculate the resistance of 30 m of this cable.

3. Calculate the resistance of 1·5 m of cable if 60 m has a resistance of 0·05 Ω.

Problems

1. A cable has a resistance of 0·4 Ω and an area of 15 mm². If the area is increased to 90 mm² what would be the resistance?

2. A cable has an area of 2 mm² and a resistance of 0·03 Ω. If the area is increased to 72 mm² what would be the resistance?

3. A cable has a cross-sectional area of 105 mm² and a resistance of 0·005 Ω. What would be the resistance if the area was only 7 mm²?

BELT-DRIVES

The creation of electricity in the alternator or dynamo can only occur because the rotor or armature is driven from the engine by the 'fan' belt (a now much misused name and more properly called the 'driving belt').

State the factors which govern the maximum torque that can be transmitted by a pulley to the belt.

....

....

....

....

....

....

Sketch a section through a belt positioned correctly on the pulley shown. Indicate one of the most important torque transmitting factors.

Draw simple sketches to show what is meant by angle of lap.

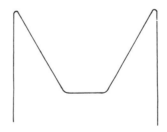

Why is it necessary to adjust an alternator's drive-belt to a greater tension than a dynamo's belt?

....

....

BATTERY CAPACITY

Battery capacity is a measure of the current that can be taken from a battery at a given discharge rate over a period of 10 or 20 h. The units for battery capacity are ....

This capacity is obtained by discharging a fully charged battery at a constant rate until the terminal voltage drops to 1.8 V per cell.

E.g. A 66 A/h battery, measured at a 20-h rate, will discharge at (66/20) 3.3 A for 20 h before the cells' voltages drop below 1.8 V.

Problem

1. Calculate the capacity of a battery at a 20-h rate when the discharge current is 4.5 A.

3. A 20 A/h battery has a capacity of 75 A/h.
   Calculate the discharge current.

4. A 10 A/h battery has a capacity of 70 A/h.
   Calculate the discharge current.

2. Calculate the capacity of a battery at a 10-h rate when the discharge current is 9.5 A.

Upon what is the capacity of a battery primarily dependent?

....

....

What can reduce the capacity of a battery?

....

....

RATE OF DISCHARGE

If a battery is discharged slowly its capacity will be greater than if it is discharged rapidly.

The capacity at the 20-h rate can be obtained (approximately) by multiplying the capacity at a 10-h rate by 8 and dividing by 7. E.g. A 60 A/h battery measured at a 10-h rate would have a capacity of 60 × 8/7 = 68.5 A/h measured at a 20-h rate, the discharge current being 68.5/20 = 3.425 A.

Problem

1. A battery has a capacity of 100 A/h at a 10-h rate.

   Calculate its approximate capacity at a 20-h rate and state the discharge current that would achieve this capacity.

2. A battery has a capacity of 84 A/h at a 20-h rate.

   Calculate its approximate capacity at a 10-h rate and state the discharge current that would achieve this capacity.

What will be the effect on A/h rating if batteries are connected in:

1. series ....

....

....

2. parallel...

ELECTROMAGNETIC INDUCTION INVESTIGATION

Examine one effect of magnetism by moving a closed loop of wire, connected to a galvanometer, through the magnet field of a U-shaped strong permanent magnet.

Show a sketch of equipment used, naming all components.

Operation	Effect
1. Move wire in and out of magnetic field created between the ends of magnet.	
2. Hold wire and move magnet.	
3. Loop wire around one of the ends of the magnet and repeat the above operations.	

If the magnets are shown in a position similar to the electromagnetic field coils in a generator or motor, the direction of current flow can be established.

1. **GENERATOR**, when a current is induced by rotating the winding (armature) the right-hand rule may be applied.

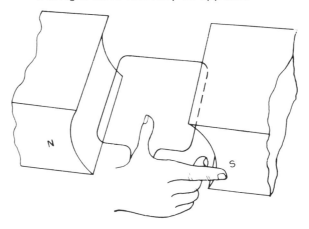

2. **ELECTRIC MOTOR**, when current is supplied to the winding and movement is created the left-hand rule may be applied.

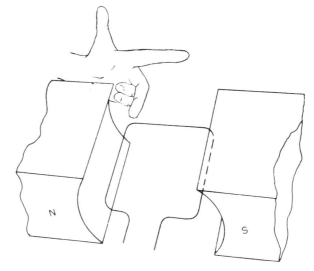

229

INVESTIGATION

Examine the effect of passing a current through a coil of wire positioned between the poles of a permanent magnet.

Show a sketch of the equipment used.

What effect does the passing of the current through the coil have?

....

....

Show and explain this action with the aid of the simple diagrams below. Complete the second diagram.

No current flowing

Current flowing

State why the current supplied to a starter motor varies with:

battery potential difference ....

....

length of cable ....

....

speed of armature ....

....

....

....

It is necessary to supply a very large amount of current to the starter motor to enable it to turn the engine from a stationary position. This causes a considerable voltage drop in the circuit. The starter motor is wired in such a way that it produces maximum torque immediately on turning (locked torque). This maximum torque is required to overcome the resistance to movement of the engine.

INVESTIGATION

Check the locked torque and locked current of a starter motor. Clamp starter motor in suitable test rig and attach torque arm.

Operate starter motor and obtain voltage drop, locked current and locked torque readings.

Starter make .... Model ....

Readings	Actual results	Manufacturer's specification
Locked current		
Locked torque		
Voltage drop		

230

PRINCIPLE OF ELECTRIC GENERATOR AND METHOD OF RECTIFICATION

In order to produce an electric current by magnetic induction three basic requirements must be fulfilled.

These are: ....

....

....

Explain with the aid of the diagrams on this page the operation of an electric generator whose current is rectified by a simple commutator.

Show the direction of current flow on each diagram.

ONE ARMATURE WINDING AND A SIMPLE COMMUTATOR

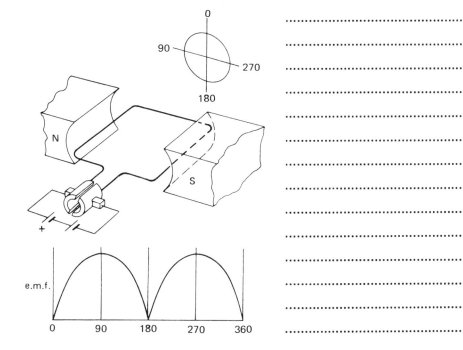

....

....

....

....

....

....

....

....

....

....

....

....

NUMBER OF WINDINGS INCREASED ON ARMATURE

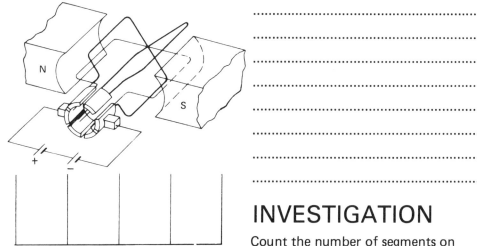

Show the e.m.f. produced

....

....

....

....

....

....

....

INVESTIGATION

Count the number of segments on an actual commutator.

No. of SEGMENTS ....

No. of LOOPS ....

CONNECTION OF FIELD WINDINGS

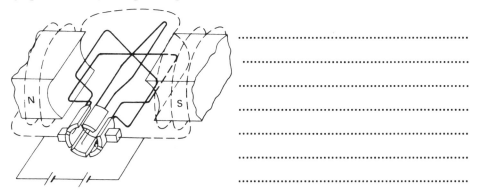

....

....

....

....

....

ALTERNATOR – PRINCIPLE OF FULL-WAVE STATIC RECTIFICATION

When the magnet in a simple alternator is revolved, one complete turn, an e.m.f. is induced in the circuit, first in one direction and then in the reverse direction.

On the diagrams below, show an alternating pulse and a rectified pulse.

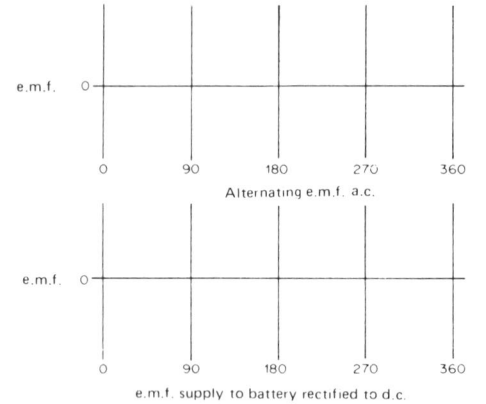

The components that allow this rectification to occur are diodes (see next page).

Show, using arrows, how diodes rectify the supply of current induced in a single coil of wire.

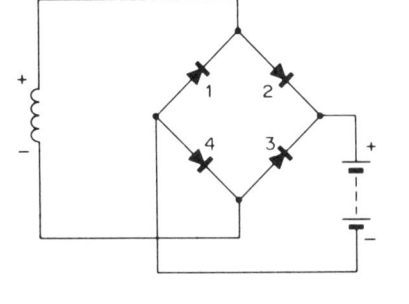
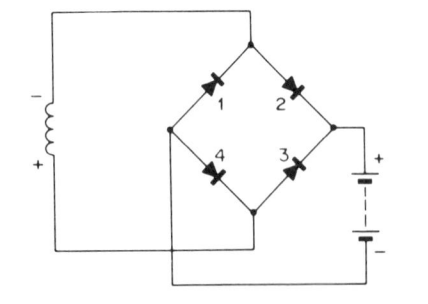

Diodes passing current Nos. .... Diodes passing current Nos. ....

The alternator stator windings are so positioned that three separate e.m.f. pulses are induced at the same time, but are slightly out of phase with one another. These windings may be wound by two different methods 'delta' and 'Wye', the latter being the more popular in this country.

In order to rectify these three pulses, six diodes are required as shown below.

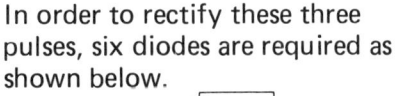
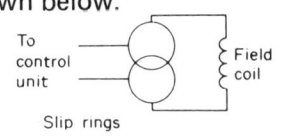

Draw a star winding arrangement connected to the diodes below.

Delta winding

Star or Wye winding

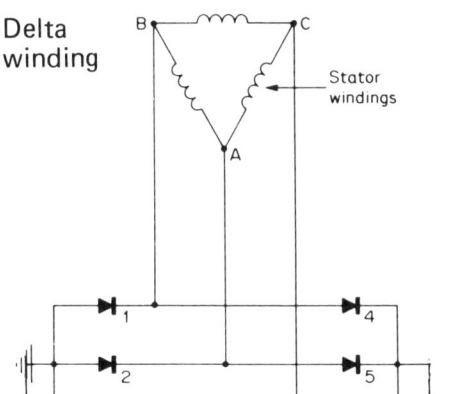
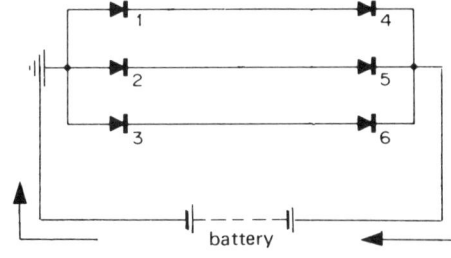
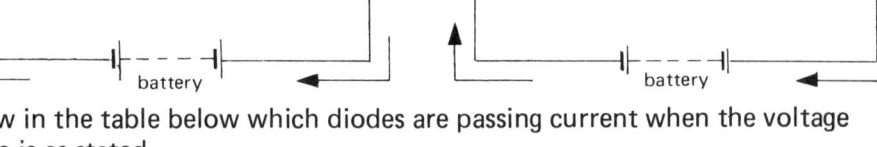

Show in the table below which diodes are passing current when the voltage pulse is as stated.

Voltage pulse	Flow transmitted by diodes	Voltage pulse	Flow transmitted by diodes
A to B	2 and 4	C to A	
B to A		B to C	
A to C		C to B	

232

FUNCTIONS OF SEMI-CONDUCTOR COMPONENTS

The electronic components listed below are all used in the alternator's electrical circuit and may be used in any other circuit that adopts electronic control.

The alternating current flow must be rectified to direct current. This is achieved by using a number of static rectifiers called diodes.

The function of a diode is to:

....

....

Show a diode's electrical symbol and indicate the direction of current flow.

What material is used in order to allow this to occur? ....

....

....

Show the electrical circuit symbol for components below and explain their basic function. Show directions of current flow.

Transistor

Avalanche diode ———— · ————

....

....

....

Surge protection diode ———— · ————

....

....

....

....

Many Continental and American charging systems still use mechanical regulators, as shown opposite, to control alternator output, while in Britain the tendency has been to use electronic regulators.

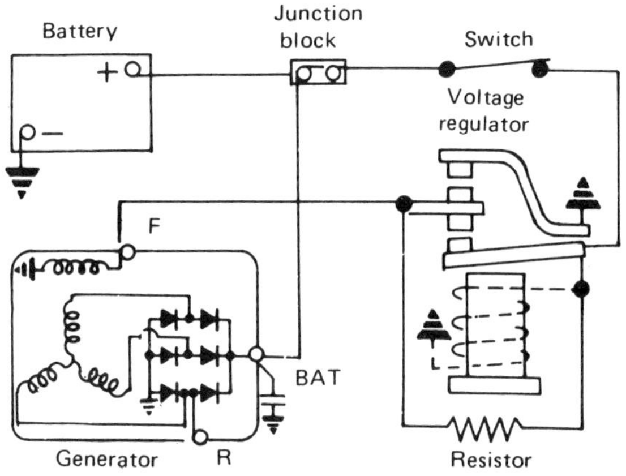

State the relative merits of:

(1) Mechanical regulators

....

....

(2) Electronic regulators

....

....

....

List the precautions required to prevent damage to semi-conductor devices.

....

....

....

....

....

....

233

INTERPRETATION OF ELECTRICAL CIRCUIT DIAGRAMS

In order to interpret vehicle wiring diagrams when seeking connections or where to test for possible faults, a knowledge of basic symbols and the method of tracing current flow round circuits must be understood.

The circuit below could represent the .... circuit of a motor vehicle

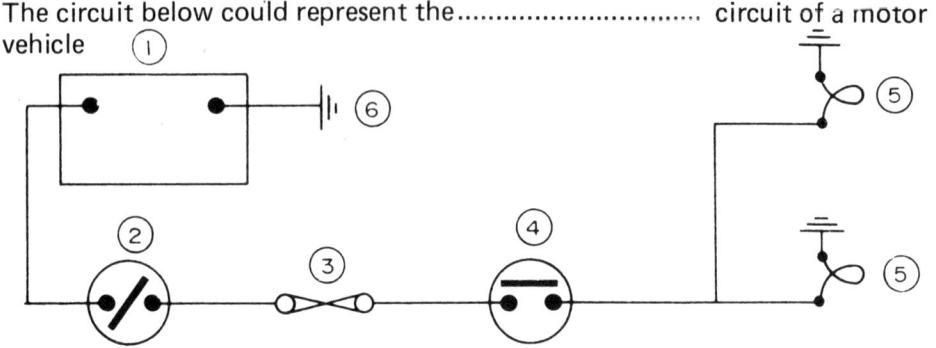

Name the numbered components in the above circuit.

1.	4.
2.	5.
3.	6.
items 5 are wired in	

The circuit below could represent the .... circuit of a motor vehicle.

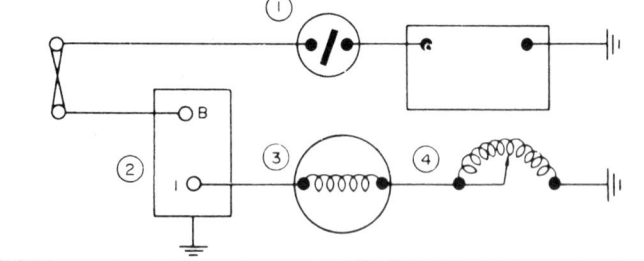

1.	3.
2.	4.

On the wiring diagrams below state if the bulbs operate when the switches are in the positions given in each table.

Use one of the following abbreviations for the bulbs' operation.

1. Not working (NW)   2. Full brilliance (FB)   3. Glowing dimly (GD)

SW$_1$	SW$_2$	Bulb
Closed	Open	
Open	Closed	
Closed	Closed	

SW$_1$	SW$_2$	Bulb$_1$	Bulb$_2$
Closed	Open		
Open	Closed		
Closed	Closed		

SW$_1$	SW$_2$	SW$_3$	Bulb$_1$	Bulb$_2$
Closed	Open	Open		
Open	Closed	Open		
Open	Open	Closed		
Open	Closed	Closed		

SW$_1$	SW$_2$	Bulb$_1$	Bulb$_2$
Closed	Open		
Open	Closed		
Closed	Closed		

SW$_1$	SW$_2$	Bulb
Closed	Open	
Open	Closed	
Closed	Closed	

Considering the final circuit what would be the effect if switch 1 were omitted and the cables reconnected?

....

234

BRITISH STANDARD RECOMMENDATIONS FOR ELECTRICAL SYMBOLS

All electrical symbols are shown in B.S. 3939 'Graphical Symbols for Electrical Power, Telecommunications and Electronics Diagrams'. They do not, however, relate specifically to motor-vehicle wiring diagrams.

In BS AU7: 1968, 'Specifications for Chart and Colour Code for Vehicle Wiring' two charts are shown which represent all the main wiring layouts contained in a motor vehicle. The symbols shown are but one representation of each electrical component, and are basically Lucas in origin.

The main aim of B.S. A.U.7 is to standardise the use of cable colours to all main switches and components. On cars produced in Britain this standardised use of colours for specific cables is widely observed.

From the B.S. colour recommendations, state for what the following colours are used.

Colour	Code	Recommended cable use
Brown		
White		
Red		
Blue		
Green		
Purple		
Black		

Since many vehicles sold in Britain are not of British origin it follows that their wiring diagrams will not adhere to British Standards. They will, however, all identify components by showing a symbol which has a very strong pictorial relationship to the actual component represented.

EXAMPLES

Draw a B.S. Symbol representing each one of the components named below.

| Ignition switch | Stop lamp switch | Dip lamp switch | Fuse unit |

| Electric motor | Gauge unit | Flasher canister |

| Starter solenoid | Ignition coil | Spark plug |

| Lamp | Capacitor |

| Resistance | Contacts | Alternator |

INTERPRETATION OF VEHICLE WIRING DIAGRAMS

Examine the wiring diagram opposite and answer questions on this and the next page.

There are three fuses shown in the fuse unit. Trace the switches from which these circuits receive their supply.

Fuse 1..

Fuse 2..

Fuse 3..

Trace the component which these fuses protect.

Fuse	Component
1.	
2.	
3.	

Which component takes its line from the non-fused connection of fuse 1?

....

This component supplies voltage to an instrument fitted to the dash panel. Name this instrument...

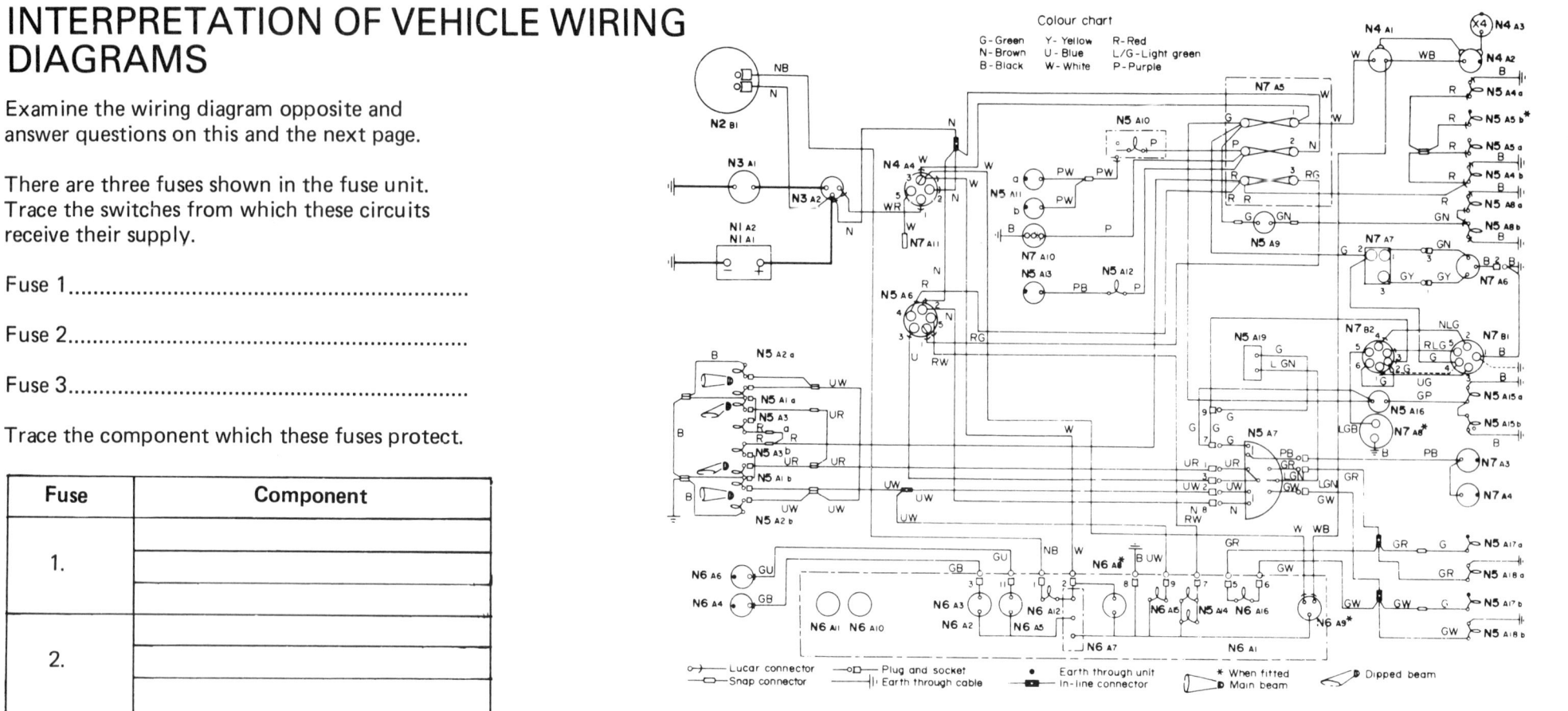

Colour chart

G - Green   Y - Yellow   R - Red
N - Brown   U - Blue   L/G - Light green
B - Black   W - White   P - Purple

o─► Lucar connector   ─o▯─ Plug and socket   ● Earth through unit   * When fitted   ◢ Dipped beam
o─▭ Snap connector   ─╫╵ Earth through cable   ▬▬▬ In-line connector   ◗ Main beam   ◢

N1	Batteries		N5A6	Light switch		N6A6	Temperature gauge bulb unit
N1A1		Battery	N5A7	Column switch		N6A7	Voltage stabiliser
N1A2			N5A8	Reverse lamp		N6A8	Voltmeter
N2	Charging System		N5A9	Reverse lamp switch		N6A9	Tachometer
N2B1	Alternator		N5A10	Interior light and switch		N6A10	Speedometer
N3	Starting System		N5A11	Courtesy light switch		N6A11	Oil gauge
N3A1	Starting motor - inertia engaged		N5A12	Glove box light		N6A12	No-charge indicator
N3A2	Starter solenoid		N5A13	Glove box light switch		N6A15	Main beam warning light
			N5A14	Panel light		N6A16	Direction indicator
N4	Ignition System		N5A15	Stop lamp			warning light
N4A1	Coil		N5A16	Stop lamp switch			
N4A2	Distributor		N5A17	Direction indicator - front	N7	Ancillaries
N4A3	Sparking plug		N5A18	Direction indicator - rear		N7A3	Horn
N4A4	Ignition switch		N5A19	Flasher unit		N7A4	Horn
				(a) Left-hand (b) Right-hand		N7A5	Fuse unit
N5	Lighting System					N7A6	Heater blower motor
N5A1	Headlamp - circular outer	N6	Instrumentation		N7A7	Blower motor switch
N5A2	Headlamp - circular inner		N6B1	Binnacle		N7A8	Screen washer pump motor
N5A3	Side lamp		N6B2	Printed circuit		N7A10	Cigarette lighter
N5A4	Tail lamp		N6B3	Fuel gauge		N7A11	Radio pickup
N5A5	Number plate lamp		N6B4	Fuel gauge tank unit		N7B1	Screenwiper
			N6B5	Temperature gauge		N7B2	Screenwiper/washer switch

The ignition switch receives, at connection 2, its current supply from the battery lead connection at the solenoid switch, the cable colour being brown (N).

State where the cables go to that are connected to the four other terminals of the switch, and indicate their colour.

Ignition switch terminal	Feed to component	Cable colour
1		
3		
5		

Reproduce below from the wiring diagram the charging circuit to include: alternator, solenoid switch, ignition switch and ignition warning light.

Note. The type of alternator used on this vehicle is a Lucas A.C.R. This type has the regulator fitted into the alternator body and therefore no external wiring is necessary.

Trace the wires from the lighting switch to the various switches and fittings. State the cable colour and if the flow is to or from the switch. Assume that the current flow is positive to negative.

Lighting switch terminal	Component connected	Cable colour
1	To fuse No. 3	Red and green
2		
3		
4		
5		

The steering column switch as shown on the diagram is shown enlarged below. Write on each cable the component to which it is connected at the other end, and state the cable's colour.

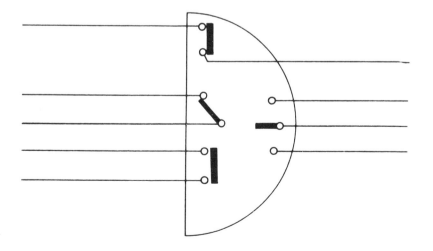

N5 A7 Column switch

237